CANCER ETIOLOGY, DIAGNOSIS AND TREATMENTS SERIES

MICRORNAS IN BREAST CANCER

CANCER ETIOLOGY, DIAGNOSIS AND TREATMENTS SERIES

Cell Apoptosis and Cancer
Albina W. Taylor (Editor)
2007. ISBN: 1-60021-506-8

Chronic Lymphocytic Leukemia Research Focus
Chadi Nabhan (Editor)
2007. ISBN: 1-60021-526-2

Cervical Cancer Research Trends
Eleanor P. Bankes (Editor)
2007. ISBN: 1-60021-648-x

Lung Cancer in Women
Varetta N. Torres (Editor)
2008. ISBN: 1-60021-659-5

Lung Cancer in Women
Varetta N. Torres (Editor)
2008. ISBN: 978-1-60692-765-6
(Online book)

Cancer Research at the Leading Edge
Ignatius K. Martakis (Editor)
2008. ISBN: 1-60021-728-1

Chronic Lymphocytic Leukemia: New Research
Inès B. Moreau (Editor)
2008. ISBN: 978-1-60456-081-7

Cancer and Stem Cells
Thomas Dittmar and Kurt S. Zander (Editors)
2008. ISBN: 978-1-60456-478-5

Cancer and Stem Cells
Thomas Dittmar and Kurt S. Zander (Editors)
2008. ISBN: 978-1-61668-044-7
(Online Book)

Cancer Prevention Research Trends
Louis Braun and Maximilian Lange (Editors)
2008. ISBN: 978-1-60456-639-0

Clinical, Genetic and Molecular Precursor Features in Colorectal Neoplasia
Kjetil Søreide and Håvard Søiland (Editors)
2008. ISBN: 978-1-60456-714-4

Drug Resistant Neoplasms
Ethan G. Verrite (Editor)
2009. ISBN: 978-1-60741-255-7

Handbook of Prostate Cancer Cell Research: Growth, Signalling and Survival
Alan T. Meridith (Editor)
2009. ISBN: 978-1-60741-954-9

Human Polyomaviruses: Molecular Mechanisms for Transformation and their Association with Cancers
Ugo Moens, Marijke Van Gheule and Mona Johannessen
2009. ISBN: 978-1-60692-812-7

Anticancer Drugs: Design, Delivery and Pharmacology
Peter Spencer and Walter Holt (Editors)
2009. ISBN: 978-1-60741-004-1

Anticancer Drugs: Design, Delivery and Pharmacology
Peter Spencer and Walter Holt (Editors)
2009. ISBN: 978-1-60876-629-1
(Online Book)

Molecular Therapy of Breast Cancer: Classicism Meets Modernity
Marc Lacroix
2009. ISBN: 978-1-60741-593-0

Molecular Therapy of Breast Cancer: Classicism Meets Modernity
Marc Lacroix
2009. ISBN: 978-1-60876-726-7
(Online Book)

Aromatase Inhibitors: Types, Mode of Actionand Indications
Jean R. Lamonte (Editor)
2009. ISBN: 978-1-60741-711-8

Cancer Biology: An Updated Global Overview
Tarek H. EL-Metwally
2009. ISBN: 978-1-60876-193-7

Aromatase Inhibitors: Types, Mode of Action and Indications
Jean R. Lamonte (Editor)
2009. ISBN: 978-1-61668-665-9

Small Cell Carcinomas: Causes, Diagnosis and Treatment
Jonathon G. Maldonado and Mikayla K. Cervantes (Editors)
2009. ISBN: 978-1-60741-787-3

Multiple Myeloma: Symptoms, Diagnosis and Treatment
Milen Georgiev and Evgeni Bachev
2009. ISBN: 978-1-60876-108-1

Viral Cancers: Cytologic Tools in Diagnosis and Management
Dilip K. Das
2010. ISBN: 978-1-60876-402-0

Nose and Viral Cancer: Etiology, Pathogenesis and Treatment
Aloisio Medeiros and Carlitos Veloso (Editors)
2010. ISBN: 978-1-60741-735-4

**Karyogamic Theory of Cancer
Cell Formation from the View of
the XXI Century**
G.Gogichadze and T.Gogichadze
2010. ISBN: 978-1-60876-386-3

Aggressive Breast Cancer
*Regina H. DeFrina
(Editor)*
2010. ISBN: 978-1-60876-881-3

**Human Papillomavirus (HPV)
Involvement in
Esophageal Carcinogenesis**
Kari Syrjänen
2010: ISBN: 978-1-60876-211-8

**New Quinolones with Potential
Anti-MRSA Activity**
Saeed Emami
2010. ISBN: 978-1-60876-736-6

PSA and Prostate Cancer
*Jake A. Saylor
and Lionel B. Michaels
(Editors)*
2010. ISBN: 978-1-60876-895-0

**Pentacyclic Triterpenes
as Promising
Agents in Cancer**
*Jorge A. R. Salvador
(Editor)*
2010. ISBN: 978-1-60876-973—5

**Breast Cancer:
Causes, Diagnosis and Treatment**
*Martin E. Romero
and Louis M. Dashek
(Editors)*
2010. ISBN: 978-1-60876-463-1

**Gastric Cancer:
Diagnosis, Early Prevention,
and Treatment**
*Victor D. Pasechnikov
(Editor)*
2010. ISBN: 978-1-61668-313-9

**New Approaches in the
Treatment of Cancer**
*Dra. Ma. Del Camen Mejia Vazquez,
and Samuel Navarro*
2010. ISBN: 978-111-61668-361-0

**Handbook of Skin Care
in Cancer Patients**
*Pierre Vereecken, Ahmad Awada
and Jules Bordet
(Editors)*
2010. ISBN: 978-1-61668-419-8

MicroRNAs in Breast Cancer
Marc Lacroix
2010. ISBN: 978-1-61668-438-9

MicroRNAs in Breast Cancer
Marc Lacroix
2010. ISBN: 978-1-61668-498-3
(Online Book)

CANCER ETIOLOGY, DIAGNOSIS AND TREATMENTS SERIES

MICRORNAS IN BREAST CANCER

MARC LACROIX

Nova Science Publishers, Inc.
New York

Copyright © 2010 by Nova Science Publishers, Inc.

All rights reserved. No part of this book may be reproduced, stored in a retrieval system or transmitted in any form or by any means: electronic, electrostatic, magnetic, tape, mechanical photocopying, recording or otherwise without the written permission of the Publisher.

For permission to use material from this book please contact us:
Telephone 631-231-7269; Fax 631-231-8175
Web Site: http://www.novapublishers.com

NOTICE TO THE READER

The Publisher has taken reasonable care in the preparation of this book, but makes no expressed or implied warranty of any kind and assumes no responsibility for any errors or omissions. No liability is assumed for incidental or consequential damages in connection with or arising out of information contained in this book. The Publisher shall not be liable for any special, consequential, or exemplary damages resulting, in whole or in part, from the readers' use of, or reliance upon, this material.

Independent verification should be sought for any data, advice or recommendations contained in this book. In addition, no responsibility is assumed by the publisher for any injury and/or damage to persons or property arising from any methods, products, instructions, ideas or otherwise contained in this publication.

This publication is designed to provide accurate and authoritative information with regard to the subject matter covered herein. It is sold with the clear understanding that the Publisher is not engaged in rendering legal or any other professional services. If legal or any other expert assistance is required, the services of a competent person should be sought. FROM A DECLARATION OF PARTICIPANTS JOINTLY ADOPTED BY A COMMITTEE OF THE AMERICAN BAR ASSOCIATION AND A COMMITTEE OF PUBLISHERS.

LIBRARY OF CONGRESS CATALOGING-IN-PUBLICATION DATA

ISBN: 978-1-61668-438-9

Available Upon Request

Published by Nova Science Publishers, Inc, ✣ New York

CONTENTS

Preface vii

Chapter 1   Introduction   1

Chapter 2   Detection and Measurement of miRNAs   5

Chapter 3   miRNA-Related Bioinformatics Tools   13

Chapter 4   miRNAs of Importance in Breast Cancer   17

Chapter 5   Clinical Potential of miRNAs in Breast Cancer   81

Chapter 6   Human miRNAs: Genes, Names, Loci, Sequences, Clusters   89

Index   112

PREFACE

MicroRNAs, or miRNAs, are a recently discovered class of small regulatory RNAs that influence the stability and translational efficiency of target messenger RNAs (mRNAs). Alterations in miRNA expression are associated with an increasing number of biological processes, including breast cancer. Some miRNAs are up-regulated in breast tumors, such as miR-21, miR-155, miR-373, and miR-520c, and appear as putative oncogenes. Other miRNAs are down-regulated, such as miR-126, and miR-145, and members of the let-7 family; functional studies support their tumor suppressor nature. miRNAs associated with estrogen receptor expression and function, such as miR-18a, miR-22, miR-181, miR-206, miR-221 and miR-222, or with HER2/neu, such as miR-125, have also been identified. Other miRNAs, such as miR-210 and miR-421 have been linked to hypoxia and drug resistance, respectively. Of peculiar interest are miRNAs (miR-200 family members, miR-205) involved in epithelial-mesenchymal transition (EMT), a process that is proposed to play an important role in cancer cell metastasis. The study of miRNAs is a rapidly developing field that could considerably change our vision of breast cancer biology.

This book offers an insight into our current knowledge of human miRNAs, with a specific interest for breast cancer. What are exactly miRNAs, how are they found, how are their target mRNAs identified, which miRNAs are of importance in breast cancer, notably from a clinical point of view ? Elements of response to these questions are brought here.

Born in Verviers (Wallonia, Belgium), Marc Lacroix has been working for more than 20 years in several academic institutions (University of Liège, Free University of Brussels, Jules Bordet Institute) and at InTextoResearch, an agency devoted to scientific information on cancer. He is specialized in estrogen receptor biology and function, metastasis and, more generally, of molecular aspects of breast cancer. He authored two books published by Nova Science Publishers: "Tumor Suppressor Genes in Breast Cancer" (2008) and "Molecular Therapy of Breast Cancer: Classicism Meets Modernity" (2009).

DEDICATION

This book is dedicated to my family

Chapter 1

INTRODUCTION

ABSTRACT

MicroRNAs, or miRNAs, are a recently discovered class of small regulatory RNAs that influence the stability and translational efficiency of target messenger RNAs (mRNAs). The biogenesis of miRNA sequences has been largely elucidated. Each miRNA possess a specific 'seed' sequence, a sequence composed of nucleotides 2 through 8 of its 5' end, which is essential for target recognition. Important involvement of miRNAs in breast cancer biology is supported by an increasing amount of papers.

MicroRNAs (miRNAs) are a class of evolutionarily conserved, non-coding RNA molecules of ~22 (between 17 and 27) nucleotides. miRNAs coding genes may be present either as single units scattered in the genome or may be organized in gene clusters, such as miR~17~92, miR~23a~27a~24-2, miR~23b~27b~24-1, miR~106a~363, miR~106b~25 and miR~222~221. Clustered miRNAs are likely transcribed together in polycistronic transcripts and usually control mRNAs with related functions. miRNAs play important roles in the regulation of target genes by binding to complementary regions of messenger transcripts to repress their translation or regulate degradation. During the last years, a number of reports have highly suggested an implication of miRNAs in human cancers [Lowery *et al.* 2008; Verghese *et al.* 2008; Adams *et al.* 2008; Iorio *et al.* 2008; Nelson and Weiss 2008; Siomi and Siomi 2009; Jinek and Doudna 2009; Bartel 2009].

The biogenesis of miRNA sequences has been largely elucidated and is described in details elsewhere (see review articles here above). Briefly, miRNAs are first transcribed as primary miRNAs (pri-miRNAs) by RNA polymerase II (Pol II). These pri-miRNAs can be hundreds to thousands nucleotides long and, like any other Pol II transcript, undergoes capping and polyadenylation. They are then cut by the RNase III, Drosha, and its cofactor, the double-stranded RNA (dsRNA)-binding protein DGCR8 ('DiGeorge syndrome critical region gene 8'; named Pasha in flies and worms) into ~70-nucleotide stem-loop ("hairpin") precursors (pre-miRNAs), which are transported to the cytoplasm by the nuclear export receptor exportin-5 (Ran-binding protein 21). Another RNAse III called Dicer, and its binding partner TRBP, then remove the loop region of the pre-miRNAs, releasing a mature miRNA: miRNA* duplex [Schwarz *et al.* 2003]. The duplex is recruited into the RNA-induced silencing complex (RISC), leading to miRNA* degradation to form single-stranded miRNAs. Important components of this complex are proteins of the Argonaute family; Ago1, Ago2, Ago3 and Ago4. Finally, this complex interferes with the translation and stability of target messenger RNAs. Each miRNA possess a 'seed' sequence, a sequence composed of nucleotides 2 through 8 of its 5' end and is essential for target recognition.

miRNA terminology can be confusing, and international guidelines have been put forward for their classification and annotation [Ambros *et al.* 2003; Griffiths-Jones *et al.* 2006]. miRNAs are assigned sequential numerical identifiers. The mature sequences are designated 'miR'. Abbreviated 3 or 4 letter prefixes designate the species of origin, such that identifiers take the form hsa-miR-"an identifying number" in *Homo sapiens* (the prefix ''mmu'' represent *Mus musculus* or mouse species, etc). The only exception to this rule in the human and mouse genomes is the let-7 family of miRNAs which have retained their originally described ''let'' prefix (e.g. hsa-let-7b, and mmu-let-7a). Unless otherwise stated, all miRNAs referred to in this review are human, and hence species relevant prefixes have been omitted. Each miRNA is also given a unique identifying number, assigned sequentially, such that identical miRNAs in different species are given the same number (they are "orthologous"). Paralogous sequences, whose mature miRNAs differ at only one or two position, are given lettered suffixes — for example, miR-27a and miR-27b in *Homo sapiens*. Distinct hairpin loci that give rise to identical mature miRNAs have numbered suffixes (e.g. miR-7-1, miR-7-2 and miR-7-3). The terms "5p" and "3p" (e.g. miR-17-5p and miR-17-3p) are used to designate miRNAs from the 5' and 3' arms, respectively.

The first miRNA (lin-4) was found in nematode *Caenorhabditis elegans* in 1993 [Lee *et al.* 1993]. It was subsequently shown to effect developmental gene expression through translational repression of the mRNA of the lin-14 gene. A second miRNA, let-7, was identified in the same nematode in 2000 [Slack *et al.* 2000]. Thereafter, miRNAs were shown to constitute a very populous class of molecules found first in worms, flies and mammals, and, more recently, in plants, green algae, viruses, and more deeply branching animals [Griffiths-Jones *et al.* 2008]. In the human genome, more than 500 mature miRNAs have been identified as yet (see Table 1 in chapter 6), and computational prediction estimates that their number could increase to >1,000 [Berezikov *et al.* 2005]. A list of miRNA clusters is provided in Table 2 of chapter 6 "Human miRNAs: genes, names, loci, sequences, clusters".

There is a great emphasis placed on miRNA research now and several miRNAs have been identified as possible biomarkers of breast cancer. However, some essential questions still remain to be fully elucidated. How are transcription and processing of miRNAs regulated? How do miRNAs find their targets? What is the mechanism of regulation?

REFERENCES

Adams BD, Guttilla IK, White BA. Involvement of microRNAs in breast cancer. Semin Reprod Med. 2008 Nov;26(6):522-36.

Ambros V, Bartel B, Bartel DP, Burge CB, Carrington JC, Chen X, Dreyfuss G, Eddy SR, Griffiths-Jones S, Marshall M, Matzke M, Ruvkun G, Tuschl T. A uniform system for microRNA annotation. RNA. 2003 Mar;9(3):277-9.

Bartel DP. MicroRNAs: target recognition and regulatory functions. *Cell.* 2009 Jan 23;136(2):215-33.

Berezikov E, Guryev V, van de Belt J, Wienholds E, Plasterk RH, Cuppen E. Phylogenetic shadowing and computational identification of human microRNA genes. *Cell.* 2005 Jan 14;120(1):21-4.

Griffiths-Jones S, Grocock RJ, van Dongen S, Bateman A, Enright AJ. miRBase: microRNA sequences, targets and gene nomenclature. *Nucleic Acids Res.* 2006 Jan 1;34(Database issue):D140-4.

Griffiths-Jones S, Saini HK, van Dongen S, Enright AJ. miRBase: tools for microRNA genomics. Nucleic Acids Res. 2008 Jan;36(Database issue):D154-8.

Iorio MV, Casalini P, Tagliabue E, Ménard S, Croce CM. MicroRNA profiling as a tool to understand prognosis, therapy response and resistance in breast cancer. *Eur. J. Cancer.* 2008 Dec;44(18):2753-9.

Jinek M, Doudna JA. A three-dimensional view of the molecular machinery of RNA interference. *Nature.* 2009 Jan 22;457(7228):405-12.

Lee RC, Feinbaum RL, Ambros V. The C. elegans heterochronic gene lin-4 encodes small RNAs with antisense complementarity to lin-14. *Cell.* 1993 Dec 3;75(5):843-54.

Lowery AJ, Miller N, McNeill RE, Kerin MJ. MicroRNAs as prognostic indicators and therapeutic targets: potential effect on breast cancer management. *Clin. Cancer Res.* 2008 Jan 15;14(2):360-5.

Nelson KM, Weiss GJ. MicroRNAs and cancer: past, present, and potential future. *Mol. Cancer Ther.* 2008 Dec;7(12):3655-60.

Schwarz DS, Hutvágner G, Du T, Xu Z, Aronin N, Zamore PD. Asymmetry in the assembly of the RNAi enzyme complex. *Cell.* 2003 Oct 17;115(2):199-208.

Siomi H, Siomi MC. On the road to reading the RNA-interference code. *Nature.* 2009 Jan 22;457(7228):396-404.

Slack FJ, Basson M, Liu Z, Ambros V, Horvitz HR, Ruvkun G. The lin-41 RBCC gene acts in the C. elegans heterochronic pathway between the let-7 regulatory RNA and the LIN-29 transcription factor. *Mol Cell.* 2000 Apr;5(4):659-69.

Verghese ET, Hanby AM, Speirs V, Hughes TA. Small is beautiful: microRNAs and breast cancer-where are we now? *J Pathol.* 2008 Jul;215(3):214-21.

Chapter 2

DETECTION AND MEASUREMENT OF MIRNAS

ABSTRACT

While Northern blotting is still considered the "gold standard" to detect miRNAs, other methods, such as cloning, in situ hybridization, or microarrays may have specific uses. Other techniques are real-time RT-PCR, bioluminescent or electrochemical detection, fluorescence correlation spectroscopy and surface-enhanced Raman spectroscopy. Most of these new methods still need standardization.

A number of methods are currently used to detect miRNAs, each with specific advantages and drawbacks.

NORTHERN BLOTTING

Considered as a "gold standard", Northern blotting is the most widely used method to detect and quantify miRNAs. Its steps involve: 1) fractionation of small RNA molecules by denaturing polyacrylamide gel electrophoresis; 2) transfer of RNA from the gel onto membrane; 3) fixation of RNA onto membrane through a cross linking procedure; 4) hybridization of the membrane with a fluorescent or radiolabeled oligonucleotide probe which is complementary to the target miRNA for hybridization to occur; 5) removal of unhybridized probe; 6) detection of the miRNA target.

Northern blotting suffers from various drawbacks: it is very time-consuming, taking over 24 h to perform, requires large amounts of RNA samples (as the miRNA prevalence in total RNA is very low) and labeled probe, is only semi quantitative, and is low-throughput (as notably compared to microarray analysis).

Several labs have developed novel Northern blotting methods with improved sensitivity for miRNA detection. Válóczi and colleagues applied locked nucleic acid (LNA)-modified oligonucleotide probes to Northern analysis. LNA is a special nucleotide whose ribose backbone is chemically modified, resulting in a greater affinity between LNA probes and target RNA. It was found that the sensitivity of this type of probe was at least tenfold higher than that of standard DNA probes [Válóczi et al. 2004]. When comparing blots with DNA oligonucleotides with blots with LNA-modified oligonucleotides, Várallyay et al. [Várallyay et al. 2008] observed that the assay time was improved from 16 h for complementary DNA (cDNA) oligonucleotide probe hybridization to 2 h for complementary LNA-modified oligonucleotide probe hybridization owing to the heightened hybridization temperature of 50°C for the LNA probes compared with 37°C for the DNA oligonucleotide probes. This heightened temperature is possible because of the enhanced thermal stability of LNAs [Cissell and Deo 2009]. The method of Pall and colleagues [Pall et al. 2007] increased the efficiency with which the miRNA was fixed onto the membrane by using soluble carbodiimide to crosslink RNA, which provided a 25–50 fold increase in miRNA detection sensitivity compared to the traditional UV crosslinking method.

CLONING

The technique of miRNA cloning is mainly used to discover new miRNAs, and the cloning frequency of miRNAs can, to some extent, reflect their relative abundance [Landgraf et al. 2007]. Several different methods, including miRNA serial analysis of gene expression (miRage) have been applied for miRNA cloning [Michael 2006], but their basic principles and procedures are similar. First, small RNA molecules are isolated by denaturing polyacrylamide gel electrophoresis, the 3′ and 5′ends of the small RNA molecules are ligated with adaptor sequences that contain restriction sites, and PCR primers are then designed based on the adaptor sequences. Then, the PCR products obtained from RT-PCR amplification are transferred into the vectors for further cloning and sequencing analysis [Li et al. 2009].

Cloning is not an accurate approach for miRNA quantification. Notably, it is characterized by low throughput and slow analysis speed, as compared to microarray.

REAL-TIME RT-PCR

Real-time RT-PCR is a solution-phase method widely used to quantify specific miRNAs in samples. However, the similar size of mature miRNAs and standard PCR primers limits the direct application of conventional RT-PCR protocols to miRNA detection. To solve this problem, a number of specific quantitative RT-PCR (qRT-PCR) techniques have been developed and optimized for miRNA detection, including real-time methods based upon reverse transcription (RT) reaction with a stem-loop primer followed by a TaqMan PCR analysis [Chen *et al.* 2005, Tang *et al.* 2006] and an end-point and real-time looped RT-PCR procedure [Varkonyi-Gasic *et al.* 2007]. RT-PCR exhibits high sensitivity due to PCR amplification, but it is limited by relatively high cost.

IN SITU HYBRIDIZATION

The major advantage of miRNA *in situ* hybridization (ISH) is that it can provide information on the location of miRNA expressed in cells or tissue as well as the miRNA abundance. As the normal DNA or RNA probes may not work well in this method owing to their poor binding affinity to target miRNA, Wienholds *et al.* introduced LNA (see "Northern blotting" here above) into the ISH probes to increase their affinity. Working on zebrafish, they established that most miRNAs were expressed in a tissue-specific manner during segmentation and the later stages, but not early in development, which demonstrated that their function was not involved in tissue fate establishment but in differentiation or maintenance of tissue identity [Wienholds *et al.* 2005].

Although ISH can precisely locate a specific miRNA within tissue, its major drawback is that it is not suitable for high-throughput profiling.

MICROARRAYS

Currently, the most widely used solid-phase, high-throughput method to detect miRNA levels is through a microarray [Calin and Croce 2006]. In this technology, miRNA oligonucleotide probes that usually have amine-modified

5′ termini are immobilized onto glass slides ("miRNA microarray"). miRNAs isolated from samples are converted into cDNA via reverse transcription using labeled primers. This cDNA is labeled with a fluorescent probe or biotinylated. Once the labeled cDNA samples have been loaded onto the microarray, the probes hybridize, and this is followed by a series of wash steps to remove any unhybridized DNA. Once it has been washed, if the hybridized cDNA is biotinylated, a streptavidin-labeled fluorophore can be added. If the cDNA has already been labeled with a fluorophore, the fluorescence intensity of each well can be measured. The emission wavelength and intensity of each well determines the level of expression of each target miRNA from the initial RNA sample. Of note, microarrays have been used to detect miRNAs in serum samples and correctly discriminate between normal and cancer patient samples [Lodes *et al.* 2009]. Although microarrays can analyze thousands of samples in a day, they are, however, very expensive to fabricate [Cissell and Deo 2009].

BIOLUMINESCENT DETECTION

This solid-phase assay has been described in [Cissell *et al.* 2008]. These authors developed a competitive oligonucleotide hybridization assay for the detection of miR-21 using the free miR-21 and Renilla luciferase (Rluc)-labeled miR-21 that competes to bind to an immobilized miR-21 complementary probe. Rluc is a bioluminescent enzyme. The assay is highly sensitive, able to detect 1 fmol of target owing to the high sensitivity of Rluc. The assay was employed successfully for the detection of miR-21 in cell extracts of human cancer cells. The hybridization assay was developed in a microplate format with a total assay time of 1.5 h and without the need for sample PCR amplification. Drawbacks of this method include the method's inability to be applied in vivo and the fact that a decrease in signal is measured to detect miRNA rather than an increase [Cissell and Deo 2009].

ELECTROCHEMICAL DETECTION

In this solid-phase miRNA analysis method, related to microarray analysis, target miRNA molecules are directly conjugated to redox active and electrocatalytic moieties, such as Ru(PD)(2)Cl(2) (PD=1,10-phenanthroline-

5,6-dione), through coordinative bonds with purine bases in the miRNA molecule. DNA probes which are complementary to the target miRNA are immobilized onto a solid support, and the miRNA is allowed to hybridize with the immobilized probe. After wash steps, the labeled miRNA is detected through measuring an increase in current. This increase in current can be correlated to the amount of miRNA target that is present in the sample. The excellent electrocatalytic activity of the Ru(PD)(2)Cl(2) towards the oxidation of hydrazine makes it possible to conduct ultrasensitive miRNA detection [Gao and Yu 2007]. The detection limit in these assays for miRNA was found to be in the high femtomolar to low picomolar range. Drawbacks of this method include the method's inability to be applied in vivo and the fact that a decrease in signal is measured to detect miRNA rather than an increase [Cissell and Deo 2009].

FLUORESCENCE CORRELATION SPECTROSCOPY

It is a solution-phase, rapid method based on two-color coincident detection [Neely *et al.* 2006]. In this assay, two organic fluorophore-labeled oligonucleotides are added to miRNA targets. The fluorophore-labeled oligonucleotides hybridize with their target nucleic acid. Unbound labeled oligonucleotides are hybridized with complementary oligonucleotides containing a quencher, decreasing the background noise. The counts of the two distinct fluorophores are measured, and the number of molecules present from these counts is determined using an algorithm. The limit of quantitation for this method has been reported as 500 fmol. This method is advantageous in its high sensitivity, the rapidity of the assay, and its ability to differentiate between single base mismatches. A drawback is the requirement of an external excitation source to excite the fluorophores, reducing the sensitivity of the assay [Cissell and Deo 2009].

SURFACE-ENHANCED RAMAN SPECTROSCOPY

In solid-phase, surface-enhanced Raman spectroscopy (SERS) [Driskell *et al.* 2008], silver nanorods are adsorbed onto a glass slide, and miRNAs are added to the nanorods. Once the miRNAs have adsorbed to the nanorods, surface enhanced Raman scattering spectra are taken. The resulting peaks are then used to identify miRNAs. This method displayed excellent reproducibility and single-nucleotide specificity, and does not require a label for detection. A drawback of this method is that spectra must be taken individually prior to running the assay to identify their representative Raman shift. Also, sequences with overlapping peaks cannot be differentiated, whereas hybridization-based methods do not have this issue [Cissell and Deo 2009].

The number of methods for miRNA analysis is rapidly increasing. However, most of these new methods still need standardization. Standardized methods reliable for medical diagnosis must be reproducible, accurate, and robust; they also must have established normal levels of a miRNA in the cell to determine whether this miRNA is upregulated or downregulated. Currently, Northern blotting is considered the "gold standard" for miRNA detection. This method, however, is time-consuming, and has limited sensitivity. Newer methods, more sensitive and able to detect simultaneously many miRNAs in a sample or a single miRNA in many samples, should be standardized by comparison of their observed miRNA levels with those of known standards as provided by Northern blotting. Once standardized, these newer, more sensitive methods can be applied to measure lower concentrations of miRNAs which may not be detectable with Northern blotting. By further developing the newer analysis methods, it will be possible to replace time-consuming methods with more time efficient ones. To develop more sensitive detection methods, it is imperative that the marker which measures the miRNA levels has a very low detection limit.

REFERENCES

Calin GA, Croce CM. MicroRNA signatures in human cancers. *Nat. Rev. Cancer.* 2006 Nov;6(11):857-66.

Chen C, Ridzon DA, Broomer AJ, Zhou Z, Lee DH, Nguyen JT, Barbisin M, Xu NL, Mahuvakar VR, Andersen MR, Lao KQ, Livak KJ, Guegler KJ. Real-time quantification of microRNAs by stem-loop RT-PCR. *Nucleic Acids Res.* 2005 Nov 27;33(20):e179.

Cissell KA, Deo SK. Trends in microRNA detection. *Anal. Bioanal. Chem.* 2009 Jun;394(4):1109-16.

Cissell KA, Rahimi Y, Shrestha S, Hunt EA, Deo SK. Bioluminescence-based detection of microRNA, miR21 in breast cancer cells. *Anal. Chem.* 2008 Apr 1;80(7):2319-25.

Driskell JD, Seto AG, Jones LP, Jokela S, Dluhy RA, Zhao YP, Tripp RA. Rapid microRNA (miRNA) detection and classification via surface-enhanced Raman spectroscopy (SERS). *Biosens. Bioelectron.* 2008 Dec 1;24(4):923-8.

Gao Z, Yu YH. Direct labeling microRNA with an electrocatalytic moiety and its application in ultrasensitive microRNA assays. *Biosens. Bioelectron.* 2007 Jan 15;22(6):933-40.

Landgraf P, Rusu M, Sheridan R, Sewer A, Iovino N, Aravin A, Pfeffer S, Rice A, Kamphorst AO, Landthaler M, Lin C, Socci ND, Hermida L, Fulci V, Chiaretti S, Foà R, Schliwka J, Fuchs U, Novosel A, Müller RU, Schermer B, Bissels U, Inman J, Phan Q, Chien M, Weir DB, Choksi R, De Vita G, Frezzetti D, Trompeter HI, Hornung V, Teng G, Hartmann G, Palkovits M, Di Lauro R, Wernet P, Macino G, Rogler CE, Nagle JW, Ju J, Papavasiliou FN, Benzing T, Lichter P, Tam W, Brownstein MJ, Bosio A, Borkhardt A, Russo JJ, Sander C, Zavolan M, Tuschl T. A mammalian microRNA expression atlas based on small RNA library sequencing. *Cell.* 2007 Jun 29;129(7):1401-14.

Li W, Duan R, Kooy F, Sherman SL, Zhou W, Jin P. Germline mutation of microRNA-125a is associated with breast cancer. *J. Med. Genet.* 2009 May;46(5):358-60.

Lodes MJ, Caraballo M, Suciu D, Munro S, Kumar A, Anderson B. Detection of cancer with serum miRNAs on an oligonucleotide microarray. PLoS One. 2009 Jul 14;4(7):e6229.

Michael MZ. Cloning microRNAs from mammalian tissues. *Methods Mol. Biol.* 2006;342:189-207.

Neely LA, Patel S, Garver J, Gallo M, Hackett M, McLaughlin S, Nadel M, Harris J, Gullans S, Rooke J (2006) *Nat Methods* 3:41–46.

Pall GS, Codony-Servat C, Byrne J, Ritchie L, Hamilton A. Carbodiimide-mediated cross-linking of RNA to nylon membranes improves the detection of siRNA, miRNA and piRNA by northern blot. *Nucleic Acids Res.* 2007;35(8):e60.

Tang F, Hajkova P, Barton SC, O'Carroll D, Lee C, Lao K, Surani MA. 220-plex microRNA expression profile of a single cell. *Nat. Protoc.* 2006;1(3):1154-9.

Válóczi A, Hornyik C, Varga N, Burgyán J, Kauppinen S, Havelda Z. Sensitive and specific detection of microRNAs by northern blot analysis using LNA-modified oligonucleotide probes. *Nucleic Acids Res.* 2004 Dec 14;32(22):e175.

Várallyay E, Burgyán J, Havelda Z. MicroRNA detection by northern blotting using locked nucleic acid probes. *Nat. Protoc.* 2008;3(2):190-6.

Varkonyi-Gasic E, Wu R, Wood M, Walton EF, Hellens RP (2007) *Plant Methods* 3:12

Wienholds E, Kloosterman WP, Miska E, Alvarez-Saavedra E, Berezikov E, de Bruijn E, Horvitz HR, Kauppinen S, Plasterk RH. MicroRNA expression in zebrafish embryonic development. *Science.* 2005 Jul 8;309(5732): 310-1.

Chapter 3

MIRNA-RELATED BIOINFORMATICS TOOLS

ABSTRACT

Research on miRNAs currently benefits of various bioinformatic tools, that have been developed to identify new miRNA genes; to search for miRNA sequence and putative target genes; to combine the compilation of miRNAs with target prediction modules; to allow functional interpretation of miRNA expression data, inference of miRNA gene regulation from mRNA transcriptomic profiles, combination of parallel mRNA and miRNA expression...

Among the algorithms that have been developed to identify new miRNA genes in humans are findMiRNA [Berezikov *et al.* 2005; Xie *et al.* 2005], ProMir [Nam *et al.* 2005], and PalGrade [Bentwich *et al.* 2005].

A series of tools have been developed to search for miRNA sequence and putative target genes (for a detailed review, see [Mendes *et al.* 2009]). Among the most-used of these tools are miRanda (http://www.microrna.org/microrna/home.do) [Enright *et al.* 2003], TargetScan(s) (http://www.targetscan.org) [Lewis *et al.* 2005], PicTar (http://www.pictar.org) [Krek *et al.* 2005], DIANA microT (http://diana.pcbi.upenn.edu/cgi-bin/micro_t.cgi) [Kiriakidou *et al.* 2004], RNAHybrid (http://bibiserv.techfak.uni-bielefeld.de/rnahybrid/) [Rehmsmeier *et al.* 2004], and MicroCosm (http://www.ebi.ac.uk/enright-srv/microcosm/ htdocs/targets/v5/).

Several databases combine the compilation of miRNAs with target prediction modules. They include miRBase (http://www.mirbase.org/) [Griffiths-Jones 2006], Argonaute (http://www.ma.uni-heidelberg.de/apps/zmf/argonaute/) [Shahi *et al.* 2006], miRNAMap (http://mirnamap.mbc.nctu.

edu.tw/) [Hsu et al. 2008], and miRGen (http://www.diana.pcbi.upenn.edu/miRGen.html) [Megraw et al. 2007].

There are also algorithms and software tools specifically developed for functional interpretation of miRNA expression data, inference of miRNA gene regulation from mRNA transcriptomic profiles, combination of parallel mRNA and miRNA expression [Barbato et al. 2009]: they include miRGator (http://genome.ewha.ac.kr/miRGator/) [Nam et al. 2008], SigTerms (http://sigterms.sourceforge.net/) [Creighton et al. 2008], TopKCEMC (http://www.stat.osu.edu/~statgen/SOFTWARE/TopKCEMC/) [Lin and Ding 2008], and GenMIR++ (http://www.psi.toronto.edu/genmir/) [Huang et al. 2007]

Additional interesting tools are miRviewer, Babelomics, M@ia, MAMI and GOmir. MiRviewer (http://people.csail.mit.edu/akiezun/miRviewer/) presents a global view of homologous miRNA genes in many species. It exhibits a comprehensive set of miRNA genes both from miRbase and candidate homologs identified using miRNAminer, an algorithm used for homologous conserved miRNA gene search in several animal species [Artzi et al. 2008]. miRviewer table shows conservation of miRNA genes, grouped by name, in addition to other information. Babelomics (http://www.babelomics.org/) [Al-Shahrour et al. 2008] and M@ia (http://maia.genouest.org/) are examples of free tools for the analysis of "-omics" data in which gene annotations include predicted microRNA. MAMI (MetA Mir:target Inference) (http://mami.med.harvard.edu/) and GOmir (http://www.bioacademy.gr/bioinformatics/projects/GOmir/) [Roubelakis et al. 2009] combine the predicted target genes data obtained from several tools including TargetScan (S), Miranda, DIANA microT, PicTar, RNAHybrid and miRtarget [Wang and Wang 2006].

REFERENCES

Al-Shahrour F, Carbonell J, Minguez P, Goetz S, Conesa A, Tárraga J, Medina I, Alloza E, Montaner D, Dopazo J. Babelomics: advanced functional profiling of transcriptomics, proteomics and genomics experiments. *Nucleic Acids Res.* 2008 Jul 1;36(Web Server issue):W341-6.

Artzi S, Kiezun A, Shomron N. miRNAminer: a tool for homologous microRNA gene search. *BMC Bioinformatics.* 2008 Jan 23;9:39.

Barbato C, Arisi I, Frizzo ME, Brandi R, Da Sacco L, Masotti A. Computational challenges in miRNA target predictions: to be or not to be a true target? *J. Biomed. Biotechnol.* 2009;2009:803069.

Bentwich I, Avniel A, Karov Y, Aharonov R, Gilad S, Barad O, Barzilai A, Einat P, Einav U, Meiri E, Sharon E, Spector Y, Bentwich Z. Identification of hundreds of conserved and nonconserved human microRNAs. *Nat Genet.* 2005 Jul;37(7):766-70.

Berezikov E, Guryev V, van de Belt J, Wienholds E, Plasterk RH, Cuppen E. Phylogenetic shadowing and computational identification of human microRNA genes. *Cell.* 2005 Jan 14;120(1):21-4.

Creighton CJ, Nagaraja AK, Hanash SM, Matzuk MM, Gunaratne PH. A bioinformatics tool for linking gene expression profiling results with public databases of microRNA target predictions. *RNA.* 2008 Nov;14(11):2290-6.

Enright AJ, John B, Gaul U, Tuschl T, Sander C, Marks DS. MicroRNA targets in Drosophila. *Genome Biol.* 2003;5 (1):R1.

Griffiths-Jones S. miRBase: the microRNA sequence database. *Methods Mol. Biol.* 2006;342:129-38.

Hsu SD, Chu CH, Tsou AP, Chen SJ, Chen HC, Hsu PW, Wong YH, Chen YH, Chen GH, Huang HD. miRNAMap 2.0: genomic maps of micro RNAs in metazoan genomes. *Nucleic Acids Res.* 2008 Jan;36 (Database issue):D165-9.

Huang JC, Babak T, Corson TW, Chua G, Khan S, Gallie BL, Hughes TR, Blencowe BJ, Frey BJ, Morris QD. Using expression profiling data to identify human microRNA targets. *Nat Methods.* 2007 Dec;4(12):1045-9.

Kiriakidou M, Nelson PT, Kouranov A, Fitziev P, Bouyioukos C, Mourelatos Z, Hatzigeorgiou A. A combined computational-experimental approach predicts human microRNA targets. *Genes Dev.* 2004 May 15;18 (10): 1165-78.

Krek A, Grün D, Poy MN, Wolf R, Rosenberg L, Epstein EJ, MacMenamin P, da Piedade I, Gunsalus KC, Stoffel M, Rajewsky N. Combinatorial microRNA target predictions. *Nat Genet.* 2005 May;37(5):495-500.

Lewis BP, Burge CB, Bartel DP. Conserved seed pairing, often flanked by adenosines, indicates that thousands of human genes are microRNA targets. *Cell.* 2005 Jan 14;120(1):15-20.

Lin S, Ding J. Integration of ranked lists via cross entropy Monte Carlo with applications to mRNA and microRNA Studies. *Biometrics.* 2009 Mar;65(1):9-18.

Megraw M, Sethupathy P, Corda B, Hatzigeorgiou AG. miRGen: a database for the study of animal microRNA genomic organization and function. *Nucleic Acids Res.* 2007 Jan;35(Database issue):D149-55.

Mendes ND, Freitas AT, Sagot MF. Current tools for the identification of miRNA genes and their targets. *Nucleic Acids Res.* 2009 May;37(8):2419-33.

Nam JW, Shin KR, Han J, Lee Y, Kim VN, Zhang BT. Human microRNA prediction through a probabilistic co-learning model of sequence and structure. *Nucleic Acids Res.* 2005 Jun 24;33(11):3570-81.

Nam S, Kim B, Shin S, Lee S. miRGator: an integrated system for functional annotation of microRNAs. *Nucleic Acids Res.* 2008 Jan;36(Database issue):D159-64.

Rehmsmeier M, Steffen P, Hochsmann M, Giegerich R. Fast and effective prediction of microRNA/target duplexes. *RNA.* 2004 Oct;10(10):1507-17.

Roubelakis MG, Zotos P, Papachristoudis G, Michalopoulos I, Pappa KI, Anagnou NP, Kossida S. Human microRNA target analysis and gene ontology clustering by GOmir, a novel stand-alone application. *BMC Bioinformatics.* 2009 Jun 16;10 Suppl 6:S20.

Shahi P, Loukianiouk S, Bohne-Lang A, Kenzelmann M, Küffer S, Maertens S, Eils R, Gröne HJ, Gretz N, Brors B. Argonaute--a database for gene regulation by mammalian microRNAs. *Nucleic Acids Res.* 2006 Jan 1;34(Database issue):D115-8.

Wang X, Wang X. Systematic identification of microRNA functions by combining target prediction and expression profiling. *Nucleic Acids Res.* 2006 Mar 20;34(5):1646-52.

Xie X, Lu J, Kulbokas EJ, Golub TR, Mootha V, Lindblad-Toh K, Lander ES, Kellis M. Systematic discovery of regulatory motifs in human promoters and 3' UTRs by comparison of several mammals. *Nature.* 2005 Mar 17;434(7031):338-45.

Chapter 4

MIRNAS OF IMPORTANCE IN BREAST CANCER

ABSTRACT

The expression of dozens of miRNAs has been shown to be modified in breast cancer compared to normal breast tissue, or between various subtypes of breast cancer. They are listed and detailed here. Some of them appear as "general" oncogenes or tumor suppressors, as their expression level is altered in several other cancer types. Other miRNAs are more specifically associated to breast cancer-restricted signaling pathways, such those downstream of estrogen receptor-α or HER2/neu.

The number of articles describing miRNA expression and action in breast cancer is rapidly increasing, from less than 10 in 2005 to more than 100 in 2009. This review intends to summarize the main data obtained as yet.

LET-7 FAMILY

The ubiquitously expressed let-7 was one of the first mammalian miRNAs to be identified. It is expressed late in mammalian embryonic development and plays an evolutionarily conserved role from *Caenorhabditis elegans* to Drosophila to mammals.

The let-7 family is comprised of 13 members (let-7-a1, a2, a3, b, c, d, e, f1, f2, g, i, and the related miR-98 and miR-202) located on 9 different chromosomes. These members correspond to 10 distinct mature sequences with, however, identical seed sequences and, very likely, overlapping sets of targets [Büssing et al. 2008]. The sequence similarity has been a challenge for assigning physiological functions to individual let-7 family members because hybridization-based techniques, such as microarray or Northern blotting, frequently cannot distinguish between highly related miRNAs. However, this might, to some degree, resemble the situation in a cell, where functional assays have thus far failed to determine any functional differences among these different family members [Büssing et al. 2008].

let-7 family members are expressed in various different adult tissues. In contrast to their expression in differentiated tissues, they are absent in human and mouse embryonic stem cells or pluripotent cell populations, and increasing expression upon differentiation seems to be a common theme. Misregulation of let-7 leads to a less differentiated cellular state and the development of cell-based diseases such as cancer. In fact, let-7 family members control multiple targets to regulate cell proliferation and differentiation as required for proper development and tumor suppression [Büssing et al. 2008; Roush and Slack 2008; Peter 2009].

let-7 targets that have been identified so far include cell cycle regulators such as CDC25A and CDK6, promoters of growth including RAS, MYC and CCND1 (coding for cyclin D1), and a number of early embryonic genes including HMGA2 (coding for a chromatin remodeling protein that activates pro-invasive and pro-metastatic genes, including snail, a transcription factor mediating epithelial to mesenchymal transition), TRIM71 and IMP-1/IGF2BP1 (coding for a RNA-binding protein that stabilizes the MYC RNA) (reviewed in [Büssing et al. 2008; Roush and Slack 2008; Peter 2009]). Additional targets in various cancers are CDC34 (an ubiquitin-conjugating enzyme that regulates Skp1/cullin/F-box (SCF)-mediated ubiquitination and cell cycle progression), ITGB3 (coding for an integrin that facilitates transforming growth factor (TGF)-β mediated induction of epithelial-mesenchymal transition in mammary epithelial cells) and PRDM1. For more information on epithelial-mesenchymal transition –or EMT-, see miR-200 family.

According to the cancer stem cell hypothesis, cancer tissue is composed of heterogeneous cell populations, that is, differentiated tumor cells, rapidly dividing amplifying cells (early precursor cells) and rare, slowly dividing tumor-initiating cells (TICs – 'cancer stem cells'). TICs have the capacity of unlimited self-renewal and can differentiate into multiple cell types, allowing them to repopulate tumors after therapy and seed metastasis to distant sites

[Charafe-Jauffret *et al.* 2008]. Work on breast cancer has provided evidence for a role of let-7 deregulation in determining tumorigenicity of breast TICs [Yu *et al.* 2008]. First, the authors enriched TICs from the human SK-BR-3 breast cancer cell (BCC) line. They did so by exploiting the fact that the TICs are resistant to chemotherapy, allowing to select these cells by passaging them in immunodeficient mice exposed to chemotherapy. The resulting TIC-enriched cell line showed increased self-renewal and a higher frequency of tumor formation in xenograft experiments compared to the parental cell line. Examination of miRNA levels in these cells showed that many miRNAs were expressed at lower levels in the derived than in the parental cell line, with the most-prominent reductions seen for let-7 family members. To demonstrate that reduced let-7 levels were functionally relevant and not simply a side effect of enhanced tumorigenicity, Yu et al. reintroduced let-7 into the TIC-enriched cell line. They observed not only a lower penetrance of self-renewal but also decreased tumor-forming capacity when the transfected cells were grafted into immunodeficient mice. Conversely, the inhibition of let-7 in the parental cell line elevated the capacity for self renewal. Importantly, reduced let-7 levels were also observed in TICs from primary patient samples, and reexpression of let-7 in these cells similarly reduced their in vitro tumorigenicity [Yu *et al.* 2008]. These results clearly demonstrate the functional importance of let-7 in regulating TIC properties in this system. In the future, it will be important to confirm this function of let-7 in immunocompetent mice by using mouse tumor cell lines.

17β-estradiol (E_2) was found to induce 21 miRNAs (including eight let-7 family members, miR-98 and miR-21) in MCF-7/p (a MCF-7 derivative that contain a bicistronic vector) BCC; these miRNAs reduced the levels of c-Myc and E2F2 proteins, which are involved in cell-cycle progression and secondary estrogen responses [Bourdeau *et al.* 2008]. This suggests a negative-regulatory loop controlling E_2 response through several miRNAs [Bhat-Nakshatri *et al.* 2009]. On the other hand

In BCC, the expression of a broad set of miRNAs, including let7 members let-7a, let-7c, let-7f, let-7g and miR-98, was decreased following E_2 treatment in an ER-dependent manner [Maillot *et al.* 2009].

Inflammation is linked clinically and epidemiologically to cancer, and NF-κB appears to play a causative role. Transient activation of Src oncoprotein was shown to mediate an epigenetic switch from immortalized breast cells to a stably transformed line that forms self-renewing mammospheres that contain cancer stem cells. Src activation triggered an inflammatory response mediated by NF-κB that directly activated Lin28 transcription and rapidly reduced let-7 microRNA levels, leading to a higher expression levels of the let-7 target IL6. IL6-mediated activation of the STAT3 transcription factor is necessary for transformation, and IL6 activates NF-κB, thereby completing a positive feedback loop. This regulatory circuit was shown to operate in other cancer cell lines, and its transcriptional signature was found in human cancer tissues. Thus, inflammation activates a positive feedback loop that maintains the epigenetic transformed state for many generations in the absence of the inducing signal [Iliopoulos et al. 2009].

LET-7A

let-7a miRNA is a founding member in the let-7 family and its down-regulation in association with over-expression of RAS and HMGA2 oncogenes has been reported. Another target of let-7a is CASP3 (coding for caspase-3, which may affect drug-induced apoptosis [Tsang and Kwok 2008].

Stem-loop sequences let-7a-2 and/or let-7a-3 were found down-regulated in breast cancer, as compared to normal tissue [Iorio et al. 2005; Volinia et al. 2006]. Their expression was relatively higher in lymph node (LN)-negative tumors, when compared to tumors with more than 10 positive LNs [Iorio et al. 2005]. let-7a expression was also reported to be higher in tumors which are estrogen receptor (ER)-positive, progesterone receptor (PR)-positive, low-grade or of the luminal A subtype [Mattie et al. 2006; Blenkiron et al. 2007]. This is in agreement with let-7a induction by E_2 in BCC [Bhat-Nakshatri et al. 2009].

According to an *in situ* hybridization analysis, let-7a was predominantly expressed in luminal epithelia and down-regulated in malignant cells [Sempere et al. 2007]. For reviews on breast tumor subtypes, see [Sorlie et al. 2003; Lacroix et al. 2004; Weigelt et al. 2008; Bertucci et al. 2009].

LET-7B, LET-7C, LET-7D

The expression of these miRNAs was found higher in tumors which are ER-positive (let-7b, let-7c), PR-positive (let-7b, let-7c), low-grade (let-7b, let-7c), of the luminal A subtype (let-7b, let-7c) and with a low proliferation index (let-7c, let-7d) [Iorio et al. 2005; Mattie et al. 2006; Blenkiron et al. 2007]. All these miRNAs were induced by E_2 in BCC [Bhat-Nakshatri et al. 2009].

LET-7E, LET-7G, LET-7I

All these miRNAs were induced by E_2 in BCC [Bhat-Nakshatri *et al.* 2009].

LET-7F

let-7f was found up-regulated in breast cancer compared with normal adjacent tissue [Yan *et al.* 2008]. However, let-7f expression was reported by others to be lower in tumors compared to normal tissue [Iorio *et al.* 2005], but relatively higher in tumors which are ER-positive, low-grade, of the luminal ("A" and "B") subtype and LN-negative [Iorio *et al.* 2005; Blenkiron *et al.* 2007]. Let-7f was induced by E_2 in BCC [Bhat-Nakshatri *et al.* 2009].

A male/female breast cancer comparison showed a different expression of 17 miRNAs between the two categories, with 4 upregulated and 13 down-regulated (including let-7f) miRNAs in male breast cancers [Fassan *et al.* 2009].

MIR-7

On one hand, miR-7 expression was associated to aggressiveness in ER-α-positive/LN-negative breast cancer. Bioinformatic analysis coupled miR-7 to cell cycle progression and chromosomal instability [Foekens *et al.* 2008].

On the other hand, most data available on miR-7 support its negative effect on tumors. In multiple human cancers, miR-7 expression was shown to be induced by a homeodomain transcription factor, HOXD10, the loss of which leads to an increased invasiveness. HOXD10 appeared to directly interact with the miR-7 chromatin. miR-7 was able to inhibit p21-activated kinase 1 (Pak1) expression, a widely up-regulated signaling kinase in multiple human cancers. Accordingly, an inverse correlation between the levels of endogenous miR-7 and Pak1 expression was noticed in cancer cells. miR-7 introduction inhibited the motility, invasiveness, anchorage-independent growth, and tumorigenic potential of highly invasive BCC [Reddy *et al.* 2008].

Bioinformatic analysis identified three miR-7 target sites in the epidermal growth factor receptor (EGFR) mRNA 3'-untranslated region (UTR). It was effectively found that miR-7 down-regulated EGFR mRNA and protein expression in various cancer cell lines (lung, breast and glioblastoma) via two of the three sites, inducing cell cycle arrest and cell death. Furthermore, miR-7 attenuated activation of Akt2 (protein kinase B) and extracellular signal-regulated kinase 1/2 (ERK1/2), two critical effectors of EGFR signaling in different cancer cell lines [Webster et al. 2008]. An inhibition of EGFR and the Akt pathway by miR-7 was further confirmed in glioblastoma [Kefas et al. 2008].

MIR-9

miR-9-1 expression was reported to be higher in breast tumors compared to normal tissue [Iorio et al. 2005]. According to the same authors, miR-9-2 expression was relatively higher in low stage tumors, while miR-9-3 expression was higher in tumors that do not show vascular invasion.

Aberrant hypermethylation was shown for miR-9-1 and four other miRNAs in 34-86% of cases in a series of 71 primary human breast cancer specimens [Lehmann et al. 2008]. Of note, miR-9-1 was also found methylated in 89% of pancreatic adenocarcinoma vs. 15% of normal pancreata [Omura et al. 2008]. Moreover, miR-9, miR-34b/c, and miR-148a were found to undergo specific hypermethylation-associated silencing in cancer cells compared with normal tissues. The involvement of miR-9, miR-34b/c, and miR-148a hypermethylation in metastasis formation was also suggested in human primary malignancies (n = 207) because it was significantly associated with the appearance of LN metastasis [Lujambio et al. 2008]. In medulloblastoma, miR-9 expression was down-regulated, while its rescued expression promoted medulloblastoma cell growth arrest and apoptosis while targeting the proproliferative truncated tyrosine kinase receptor C (TrkC) isoform [Ferretti et al. 2009]. It has been suggested that miR-9 may act as a putative tumor suppressor gene in recurrent ovarian cancer [Laios et al. 2008].

E_2 was found to repress seven microRNAs, including miR-9, in MCF-7 BCC [Bhat-Nakshatri et al. 2009].

Early exposure to xenoestrogens may predispose to breast cancer risk later in adult life. It is likely that long-lived, self-regenerating epithelial progenitor cells are more susceptible to these exposure injuries over time and transmit the injured memory through epigenetic mechanisms to their differentiated progeny. In progenitor-containing mammospheres exposed *in vitro* to a synthetic estrogen, diethylstilbestrol, expression profiling identified that, relative to control cells, 9.1% of miRNAs (82 of 898 loci) were altered. One of these miRNAs, miR-9-3, was down-regulated. This was accompanied by recruitment of DNA methyltransferase 1 that caused an aberrant increase in DNA methylation of its promoter CpG island in mammosphere-derived epithelial cells on diethylstilbestrol preexposure. As functional analyses suggest that miR-9-3 plays a role in the p53-related apoptotic pathway, it is therefore likely that epigenetic silencing of this gene may reduce this cellular function and promote the proliferation of BCC [Hsu *et al.* 2009].

MIR-10

A study revealed that miR-10a expression was relatively higher in luminal breast tumors compared to basal and ERBB2 (HER2/neu)-overexpressing lesions, and in ER-positive breast cancers [Blenkiron *et al.* 2007]. miR-10a has been shown to repress expression of the homeobox gene HOXD4 at the transcriptional level in BCC [Tan *et al.* 2009]. Homeobox genes are a group of evolutionarily conserved members that regulate animal morphological diversity at the organismal and evolutionary level.

miR-10b was identified as one of the most consistently down-regulated miRNAs in breast cancer compared to normal tissue. Its expression was relatively higher in tumors in which vascular invasion was absent [Iorio *et al.* 2005]. In a study of miRNA expression signature of various solid tumors, miR-10b was one of the 12 miRNAs down-regulated in breast cancers compared to normal tissue [Volinia *et al.* 2006].

Rather paradoxically, the level of miR-10b expression in primary breast carcinomas was found to correlate with clinical progression [Ma *et al.* 2007]. Indeed, miR-10b was highly expressed in metastatic BCC and able to positively regulate cell migration and invasion. Overexpression of miR-10b in otherwise non-metastatic breast tumors initiated robust invasion and metastasis. Expression of miR-10b was induced by the transcription factor Twist, which binds directly to the putative promoter of mir-10b. The miR-10b induced by Twist inhibited translation of the messenger RNA encoding homeobox HOXD10, resulting in increased expression of a well-characterized pro-metastatic gene, RHOC [Ma *et al.* 2007]. The RhoC GTPase (coded by RHOC) serves as a transforming oncogene by regulation of genes involved in the cell cycle, secretion of angiogenic factors, and activity of insulin-like growth factor. It is nearly always overexpressed in the most aggressive form of breast cancer, inflammatory breast cancer and, more generally, is associated with biologically aggressive carcinomas of the breast [Kleer *et al.* 2005].

While the above data associated high miR-10b expression level with metastasis or poor prognosis, this was not supported by a study including a higher number of patients [Gee *et al.* 2008].

MIR-15 AND MIR-16

miR-15a and miR-16-1 may act as tumor suppressors by targeting the oncogene BCL2 and inducing apoptosis [Cimmino *et al.* 2005]. These miRNAs form a cluster at the chromosomal region 13q14. Deletion or down-regulation of miR-15a and miR-16-1 has been observed in pituitary adenomas [Bottoni *et al.* 2005], chronic lymphocytic leukemia [Calin *et al.* 2008], gastric cancer [Xia *et al.* 2008], advanced prostate tumors [Bonci *et al.* 2008], ACTH-secreting pituitary tumors [Amaral *et al.* 2009]…

A male/female breast cancer comparison showed a different expression of 17 miRNAs between the two categories, with 4 upregulated and 13 down-regulated (including miR-15b and miR-16) miRNAs in male breast cancers [Fassan *et al.* 2009].

MIR-17 AND THE MIR~17~92 CLUSTER

The miR~17~92 cluster is a polycistronic gene coding for six miRNAs (miR-17, miR-18a, miR-19a, miR-20a, miR-19b-1 and miR-19b-2), obtained from the processing of a single transcript sequence, pri-mir-17-92. The cluster has a very complex history [Tanzer and Stadler 2004]. It resides in intron 3 of the C13orf25 gene at 13q31.3, a region that undergoes loss of heterozygosity (LOH) in breast cancer.

The miR-17 precursor was shown to encode 2 mature miRNAs, miR-17-3p and miR-17-5p, from its 3-prime [Lagos-Quintana *et al.* 2001] and 5-prime ends [Mourelatos *et al.* 2002], respectively. It was later found that miR-17-5p was the predominant one [Landgraf *et al.* 2007], and it was renamed miR-17 (while the minor sequence miR-17-3p was renamed miR-17*). Other mature members of the miR~17~92 cluster are miR-18, miR-19a, miR-19b, miR-20a, and miR-92.

Ancient duplications of the cluster have given rise to 2 paralogues in mammals: clusters mir~106b~25 and mir~106a~363, coding for the pri-mir~17~92 paralogue transcripts pri-mir~106b~25 and pri-mir~106a~363, respectively.

Overexpression of the miR~17~92 locus has been noted in many cancer types, including breast tumors. Notably, in a study of 540 samples including lung, breast, stomach, prostate, colon, and pancreatic tumors, miR-17 was overexpressed in breast, colon, lung, pancreas, prostate [Volinia et al. 2006]. It appears that miRNAs encoded by the cluster may act as oncogenes. Expression of these miRNAs promotes cell proliferation, suppresses apoptosis of cancer cells, and induces tumor angiogenesis [Mendell 2008].

The miR~17~92 locus is known to be regulated by the MYC oncogene and the E2F family of transcription factors [Dews et al. 2006; Sylvestre et al. 2007; Woods et al. 2007]. In fact, miR~17~92, E2F, and Myc are involved in a complex system of feedback loops [Aguda et al. 2008].

Although miRNAs are generally predicted to target hundreds of genes [Krek et al. 2005; Lewis et al. 2005], experimental evidence of miRNA-mRNA interactions from the miR~17~92 cluster has been limited to a few key components. CDKN1B and the hypoxia-inducible HIF1A are regulated by the miR~17~92 cluster, E2F1, NCOA3, RBL2 are targets of miR-17; PCAF, RUNX1, and TGFBR2 are targets of both miR-17 and miR-20a; additional miR-17 targets [Cloonan et al. 2008] are GAB1, MAPK9, MYCN, PKD1, PKD2, RBL1, TSG101, BCL2L11, PCAF, E2F1, RBL2, APP, CDKN1A, EREG, CUL3, MAPK9, NR4A3, of which many are known to be involved in tumorigenesis and/or transformation of cells. Of note, while genes such as TSG101, RBL1 and MAPK9 are inhibitors of cell proliferation, others such as MYCN, NCOA3, and NR4A3 are known promoters of cell proliferation. Thus, it is probable that miR-17 miRNA is able to act as both an oncogene and a tumor suppressor in different cellular contexts [Cloonan et al. 2008].

A cyclin D1/(miR-17 and miR-20a) regulatory feedback loop was identified in breast tumors and cell lines. Levels of these miRNAs were inversely correlated to cyclin D1 abundance. Cyclin D1 appeared to induce miR-17/miR-20a. In turn, miR-17/20 expression limited the proliferative function of cyclin D1, by negatively regulating cyclin D1 translation via a conserved 3'-UTR miRNA-binding site [Yu et al. 2008].

As mentioned above, miR-17 may down-regulate the expression of NCOA3 (nuclear receptor coactivator 3, also known as SRC-3, AIB1, p/CIP, RAC3, ACTR, and TRAM1), primarily through translational inhibition. NCOA3 interacts with nuclear receptors, notably ER-α, and certain other transcription factors, recruits histone acetyltransferases and methyltransferases for chromatin remodeling and facilitates target gene transcription. NCOA3 down-regulation resulted in decreased ER-α-mediated, as well as ER-α-independent, gene expression and decreased proliferation of BCC. miR-17 also completely abrogated the insulin-like growth factor 1-mediated, anchorage-independent growth of BCC. Thus, miR-17 may have a role as a tumor suppressor in BCC [Hossain et al. 2006].

The ability of miR-18a and miR-18b to potently regulate ER-α expression by directly targeting the 3'-UTR of ER-α mRNA was shown by reporter assays [Leivonen et al. 2009].

E2F2, E_2-inducible transcription factor involved in secondary estrogen responses [Bourdeau et al. 2008], is a target of let7/miR-98 family members, but also of miR-20a [Bhat-Nakshatri et al. 2009]. Of interest, E2F2 directly bind the promoter of the miR~17~92 (containing MIR20A), activating its transcription. This suggests an autoregulatory feedback loop between E2F2 factors and miRNAs from the miR~17~92, at least miR-20a.

MIR-20

For miR-20a, see above information on miR-17 and the miR~17~92 cluster, notably reference [Yu et al. 2008]

miR-20b is member of the miR~106a~363 cluster (see below) and is obtained from the processing of the single sequence, pri-mir~106a~363. Pri-mir~106a~363 was up-regulated by E_2 in MCF-7 BCC. In turn, miR-20b was found to down-regulate ER-α expression in these cells, indicating a negative autoregulatory feedback loop [Castellano et al. 2009].

MIR-21

miR-21 was described as one of the most consistently up-regulated miRNAs in breast cancer compared to normal tissue. Its expression was relatively higher in high clinical stage tumors [Iorio et al. 2005]. From a large-scale miRnome analysis on 540 samples including lung, breast, stomach, prostate, colon, and pancreatic tumors, miR-21 appeared as the sole miRNA overexpressed in the six tumor types considered, strongly supporting its general role as an oncogene in cancer [Volinia et al. 2006]. In an analysis involving 435 miRNA oligonucleotide probes, miR-21 appeared as the most significantly up-regulated miRNA in breast cancer compared to normal adjacent tumor tissues [Yan et al. 2008].

In addition to confirming the observations of [Iorio et al. 2005] and [Volinia et al. 2006], a study examined the behavior of MCF-7 BCC after transfection with an anti-miR-21 oligonucleotide. In these cells, both cell growth *in vitro* and tumor growth in the xenograft mouse model were suppressed. Furthermore, this anti-miR-21-mediated cell growth inhibition was associated with increased apoptosis and decreased cell proliferation, which could be in part owing to down-regulation of the antiapoptotic BCL-2 in anti-miR-21-treated tumor cells. Together, these results suggest that miR-21 functions as an oncogene and modulates tumorigenesis through regulation of genes such as BCL-2 and thus, it may serve as a novel therapeutic target [Si et al. 2007].

miR-21 overexpression in breast cancer has been associated with features of aggressive disease, such as high tumor grade, negative hormone receptor status. In BCC, miR-21 expression has been correlated with HER2/neu up-regulation and is functionally involved in HER2/neu-induced cell invasion [Huang et al. 2009]. High miR-21 was also positively correlated with TGF-β1. No associations were found between patient survival and miR-21 expression among all patients, but high miR-21 was associated with poor disease-free survival in early stage patients despite no value for prognosis [Qian et al. 2009]. Another study of 113 breast cancer cases showed that high level expression of miR-21 was significantly correlated with advanced clinical stage, LN metastasis, and shortened survival of the patients. Multivariate Cox regression analysis revealed this prognostic impact to be independent of disease stage and histological grade, suggesting that miR-21 might serve as a molecular prognostic marker for breast cancer and disease progression [Yan et al. 2008].

The fact that high miR-21 was positively correlated with TGF-β1 in breast tumors [Qian et al. 2009] is not surprising since miR-21 has been shown to be up-regulated by members of the TGF-β and bone morphogenic protein (BMP) family of growth factors and their specific SMAD signal transducers [Davis et al. 2008; Hoover and Kubalak 2008].

miR-21 has been shown to inhibit the expression of the tumor suppressors tropomyosin 1 (TPM1) and programmed cell death 4 (PDCD4), and that of maspin (SERPINB5), which has been implicated in invasion and metastasis [Zhu et al. 2007; Frankel et al. 2008; Zhu et al. 2008; Lu et al. 2008; Bourguignon et al. 2009; Huang et al. 2009].

A miRNA microarray analysis of MCF-7 BCC lines that are either sensitive or resistant to tamoxifen showed marked down-regulation of miR-21 miRNA in resistant cells compared with parental MCF-7 cells [Miller et al. 2008]. E_2 was shown to inhibit miR-21 expression in MCF-7 BCC, and this effect was prevented by the antiestrogens 4-hydroxytamoxifen (4-OHT) and ICI 182 780, and also by an ER-α directed siRNA, indicating that the inhibition is ER-α-mediated. 4-OHT increased miR-21 expression. E_2 increased luciferase activity from reporters containing the miR-21 recognition elements from the 3'-UTR of miR-21 target genes, corroborating that E_2 represses miR-21 expression resulting in a loss of target gene suppression. The E_2-mediated decrease in miR-21 correlated with increased protein expression of endogenous miR-21-targets PDCD4, PTEN and BCL2. siRNA knockdown of ER-α blocked the E_2-induced increase in PDCD4, PTEN and BCL2. Transfection of MCF-7 BCC with antisense to miR-21 mimicked the E_2-induced increase in PDCD4, PTEN and BCL2. Thus, E_2 may repress the expression of miR-21 by activating ER-α in MCF-7 cells [Wickramasinghe et al. 2009].

E_2 was found to induce 21 miRNAs (including eight let-7 family members, miR-98 and miR-21) in MCF-7/p BCC; these miRNAs reduced the levels of c-Myc and E2F2 proteins [Bhat-Nakshatri et al. 2009]. However, in MCF-7 BCC, the expression of a broad set of miRNAs, including miR-21, was decreased following E_2 treatment in an ER-α-dependent manner [Maillot et al. 2009]. A possible explanation for this discrepancy is that Bhat-Nakshatri and colleagues used a derivative of MCF-7 cells (MCF-7/p) that contain a bicistronic vector, while Maillot and colleagues, as well as Wickramasinghe and colleagues used the parental MCF-7 cell line.

To obtain insight into miRNA deregulation in breast cancer, an *in situ* hybridization method was implemented to reveal the spatial distribution of miRNA expression in archived formalin-fixed, paraffin-embedded specimens representing normal and tumor tissue from >100 breast cancer cases. miR-21 was detected predominantly within luminal epithelial cells in normal tissue, and its expression of was frequently increased in malignant cells [Sempere et al. 2008].

In MCF-7 breast tumor cells, hyaluronan binding to CD44 (a primary HA receptor) promotes protein kinase C (PKC)-ε activation which, in turn, increases the phosphorylation of the stem cell marker, Nanog. Phosphorylated Nanog is then translocated from the cytosol to the nucleus and becomes associated with RNase III DROSHA and the RNA helicase, p68. This process leads to miR-21 production, and the tumor suppressor protein, PDCD4 reduction. All of these events contribute to upregulation of inhibitors of apoptosis proteins (IAPs) and the multidrug resistant protein (MDR1) resulting in anti-apoptosis and chemotherapy resistance. To evaluate the role of miR-21 in oncogenesis and chemoresistance, MCF-7 cells were transfected with a specific anti-miR-21 inhibitor in order to silence miR-21 expression and inhibit its target functions. This inhibitor enhanced PDCD4 expression and binding to the translation initiation factor eIF4A, leading to translational inhibition. The anti-miR-21 inhibitor also blocked HA/CD44-mediated tumor cell behaviors in MCF-7 cells [Bourguignon *et al.* 2009].

Flat epithelial atypia (FEA) of the breast is characterized by a few layers of mildly atypical luminal epithelial cells. Genetic changes found in ductal carcinoma *in situ* (DCIS) and invasive ductal breast cancer (IDC) are also found in FEA, albeit at a lower concentration. Immunohistochemistry studies indicated that upregulation of miR-21 from normal ductal epithelial cells of the breast to FEA, DCIS and IDC parallels morphologically defined carcinogenesis [Qi *et al.* 2009].

MIR-22

miR-22 was found to repress ER-α expression by directly targeting the ER-α mRNA 3' UTR. Of the three predicted miR-22 target sites in the 3' UTR, the evolutionarily conserved one is the primary target. miR-22 overexpression led to a reduction of ER-α level, at least in part by inducing mRNA degradation, and compromised estrogen signaling, as exemplified by its inhibitory impact on the ER-α-dependent proliferation of BCC [Pandey and Picard 2009].

MIR~23A~27A~24-2 AND MIR~23B~27B~24-1 CLUSTERS

These homologous clusters are located at 19p13.12 and 9q22.32, respectively. The first includes MIR23A, MIR27A and MIR24-2; the second is composed of MIR23B, MIR27B and MIR24-1.

MiR-23, MiR-24, MiR-27

According to microarray-based expression profiles performed on cancer cell lines, a specific spectrum of miRNAs (including miR-23a, miR-23b, miR-24-1, miR-27a) was induced in hypoxic conditions [Kulshreshtha et al. 2007].

A miRNA microarray analysis of MCF-7 BCC lines that are either sensitive (parental) or resistant to tamoxifen showed marked down-regulation of miR-23, miR-24 and miR-27 miRNA in resistant cells compared with parental MCF-7 cells [Miller et al. 2008].

miR-24 expression was relatively higher in ER-positive and PR-positive breast tumors [Mattie et al. 2006].

miR-27a was reported to target the transcription factor ZBTB10/RINZF, which is a putative suppressor of specificity protein (Sp). Overexpression of Sp by miR-27a contributes to the increased expression of Sp-dependent survival and angiogenic genes, including survivin, vascular endothelial growth factor (VEGF), and VEGF receptor 1 (VEGFR1). Thus, the oncogenic activity of miR-27a in BCC is due, in part, to suppression of ZBTB10 [Mertens-Talcott et al. 2007].

The FOXO1 transcription factor orchestrates the regulation of genes involved in the apoptotic response, cell cycle checkpoints, and cellular metabolism. FOXO1 is a putative tumor suppressor. FOXO1 mRNA was shown to be down-regulated in breast tumor samples as compared with normal breast tissue. Silencing of the miRNA processing enzymes, Drosha and Dicer, led to an increase in FOXO1 expression. Functional and specific miRNA target sites in the FOXO1 3'-UTR for miR-27a, miR-96, and miR-182 were identified; these miRNAs have previously been linked to oncogenic transformation. In MCF-7 BCC, in which the level of FOXO1 protein is very low, all three miRNAs were observed to be highly expressed. Antisense inhibitors to each of these miRNAs led to a significant increase in endogenous FOXO1 expression and to a decrease in cell number in a manner that was blocked by FOXO1 siRNA. Overexpression of FOXO1 resulted in decreased cell viability because of inhibition of cell cycle traverse and induction of cell death. Thus, FOXO1 regulation, and targeting of FOXO1 by miRNAs may contribute to transformation or maintenance of an oncogenic state in BCC [Guttilla and White 2009].

It was shown that miR-27b and its putative target gene, ST14 (suppressor of tumorigenicity 14), had inverse expression pattern in BCC. The 3'-UTR of ST14 contains a regulatory element for miR-27b, and an antisense miR-27b enhanced ST14 expression in BCC. Furthermore, antagomir of miR-27b suppressed cell invasion, whereas pre-miR-27b stimulated invasion in BCC. In addition, ST14 reduced cell proliferation as well as cell migration and invasion. Analysis of human breast tumors revealed that miR-27b expression increases during cancer progression, paralleling a decrease in ST14 expression. However, introduction of miR-27b into ST14-expressing cells did not suppress the effect on cell growth. These findings suggest that ST14 plays an important role in several biological processes, and some effects are not completely dependent on miR-27b regulation [Wang *et al.* 2009a].

E_2 was found to induce 21 miRNAs (including miR-23a) and to repress 7 miRNAs (including miR-27a and miR-27b) in MCF-7/p BCC [Bhat-Nakshatri *et al.* 2009]. On the other hand, the expression of a broad set of miRNAs, including miR-23a, miR-23b, miR-24, miR-27a and miR-27b, was decreased in BCC following E_2 treatment in an ER-dependent manner [Maillot *et al.* 2009].

MIR-25

A male/female breast cancer comparison showed a different expression of 17 miRNAs between the two categories, with 4 upregulated and 13 down-regulated (including miR-25) miRNAs in male breast cancers [Fassan *et al.* 2009].

MIR-26

miR-26a and miR-26b expression levels were relatively higher in ER-positive and PR-positive (for miR-26a) tumors and in lesions with a low proliferation index [Iorio *et al.* 2005; Mattie *et al.* 2006]. Microarray-based expression profiles showed that a specific spectrum of miRNAs (including miR-26a-1, miR-26a-2 and miR-26b) was induced in cancer cells in response to low oxygen, maybe via a hypoxia-inducible-factor-dependent mechanism [Kulshreshtha *et al.* 2007].

Among the identified targets of miR-26a are EZH2 and SMAD1 [Sander *et al.* 2009; Luzi *et al.* 2008]. EZH2 encodes the polycomb complex protein Enhancer of Zeste 2. This protein is a transcriptional repressor critically involved in controlling cellular memory and has been linked to tumorigenesis in multiple organs. In breast cancer, EZH2 represses E-cadherin and ER-α expression, and has been identified as a potential marker that distinguishes aggressive breast cancer from indolent one [Hwang et al. 2008; Tonini et al. 2008; Cao et al. 2008]. SMAD1 is a mediator of TGF-β and bone morphogenetic protein (BMP) activity in BCC, and thus contribute to breast cancer progression and dedifferentiation in ER-positive breast cancer [Helms et al. 2005].

In BCC, the expression of miR-26a and miR-26b was decreased following E_2 treatment in an ER-α -dependent manner. In addition, a transcriptome analysis revealed that the E_2-repressed miR-26a and miR-181a regulate many genes associated with cell growth and proliferation, including the progesterone receptor gene, a key actor in estrogen signaling [Maillot et al. 2009].

MIR-29

miRNA-29a, miRNA29b-1, miR-29b-2 and miR-29c have been found up-regulated in breast cancer compared to normal tissue [Volinia et al. 2006; Yan et al. 2008, for miR-29b-2 and miR-29c]. The expression level of some or all of them was relatively higher in ER-positive and PR-positive tumors, as well as in lesions without vascular invasion [Iorio et al. 2005; Mattie et al. 2006].

MIR-30

Seven mature miR-30 family members exist: miR-30a-3p, miR-30a-5p, miR-30b, miR-30c, miR-30d, miR-30e-3p, miR-30e-5p. miR-30a-3p and miR-30c were shown to be less expressed in breast tumors compared to normal tissue [Volinia et al. 2006, Yan et al. 2008]. Among tumors, the expression of miR-30a-3p, miR-30a-5p, miR-30b, miR-30c, miR-30d was higher in ER-positive and PR-positive lesions, miR-30a-5p was higher in tumors with low proliferative index, miR-30a-3p and miR-30a-5p were relatively overexpressed in luminal tumors [Iorio et al. 2005; Mattie et al. 2006, Blenkiron et al. 2007].

E_2 was able to induce miR-30b in BCC [Bhat-Nakshatri et al. 2009].

In BCC, miR-30e was part of a miRNA signature corresponding with HER2/neu receptor status identified by stepwise artificial neural networks analysis [Lowery et al. 2009]

MIR-31

miR-31 has been described either as up-regulated [Volinia *et al.* 2006] or down-regulated in breast cancer compared to normal tissue [Yan *et al.* 2008].

miR-31 was inversely correlated with metastasis in human breast cancer patients. Overexpression of miR-31 in otherwise-aggressive breast tumor cells suppressed metastasis. miR-31-mediated inhibition of several steps of metastasis, including local invasion, extravasation or initial survival at a distant site, and metastatic colonization was achieved via coordinate repression of a cohort of metastasis-promoting genes, including RhoA. Indeed, RhoA re-expression partially reverses miR-31-imposed metastasis suppression. These findings indicate that miR-31 uses multiple mechanisms to oppose metastasis [Valastyan *et al.* 2009a].

It was shown that inhibition of breast cancer metastasis by miR-31, predicted to modulate >200 mRNAs, could be entirely explained by miR-31's pleiotropic regulation of three targets. Thus, concurrent re-expression of integrin-alpha5, radixin, and RhoA abrogated miR-31-imposed metastasis suppression. These effectors influenced distinct steps of the metastatic process [Valastyan et al. 2009b].

MIR-34

Up-regulation of miR-34 has been shown in breast tumors compared to normal tissue [Iorio *et al.* 2005]. This contrasts with the fact that miR-34 family members are often down-regulated in cancers [He *et al.* 2007]. On the other hand, CpG methylation of the miR-34a promoter was detected in breast (6/24; 25%) [Lodygin *et al.* 2008].

Several studies have found the various miR-34's to be direct transcriptional targets of p53. Activation of miR-34 members can recapitulate elements of p53 activity, including induction of cell-cycle arrest and promotion of apoptosis, and loss of miR-34 can impair p53-mediated cell death [Hermeking 2007]. Since these miRNAs may regulate the levels of hundreds of different proteins, these findings add a new, challenging layer of complexity to the p53 network (reviewed in [Lacroix *et al.* 2006])

In BCC, exogenous addition of miR-34 altered cell survival post-radiation [Kato *et al.* 2002].

MIR-96

See miR~183~96~182 cluster

MIR-98

miR-98, a member of the let-7 family of miRNAs [Büssing et al. 2008], was found up-regulated in breast cancer compared to normal adjacent tissues [Yan et al. 2008].

E_2 was found to induce 21 miRNAs (including eight let-7 family members, miR-98 and miR-21) in MCF-7 BCC; these miRNAs reduced the levels of c-Myc and E2F2 proteins [Bhat-Nakshatri et al. 2009].

Both Let-7 family members and miR-98 have the same seed sequence and target the same mRNAs. Ras family oncogenes and HMGA2 are the well-characterized targets of Let-7 (in [Bhat-Nakshatri et al. 2009])

Based on TargetScan and miRGen analyses, E2F1 and E2F2, E2-inducible transcription factors involved in secondary estrogen responses (Bourdeau et al. 2008), are predicted targets of miR-205 and Let7/miR-98, respectively. E2F1 and E2F2 are also targets of miR20 (in [Bhat-Nakshatri et al. 2009]).

In BCC, the expression of a broad set of miRNAs, including let-7 members and miR-98, was decreased following E_2 treatment in an ER-dependent manner [Maillot et al. 2009].

MIR-100

A male/female breast cancer comparison showed a different expression of 17 miRNAs between the two categories, with 4 upregulated and 13 down-regulated (including miR-100) miRNAs in male breast cancers [Fassan et al. 2009].

MIR-103 AND MIR-107

Microarray-based expression profiles revealed that a specific spectrum of miRNAs (including miR-103-1, miR-103-2, miR-107) was induced in response to low oxygen, at least some via a hypoxia-inducible-factor-dependent mechanism [Kulshreshtha et al. 2007]

miR-103 and miR-107 were predicted by bioinformatics to regulate multiple mRNA targets in metabolic pathways that involve cellular Acetyl-CoA and lipid levels [Wilfred et al. 2007].

E_2 was found to induce 21 miRNAs (including miR-103 and miR-107) in MCF-7 BCC [Bhat-Nakshatri et al. 2009].

MIR~106A~363 AND MIR~106B~25 CLUSTERS

These clusters are paralogues of the miR~17~92 cluster [Tanzer and Stadler 2004]. The first map to Xq26.2 and consists of six miRNAs, miR-106a (MIR106A), miR-18b (MIR18B), miR-20b (MIR20B), miR-19b-2 (MIR19B2), miR-92a-2 (MIR92A2) and miR-363 (MIR363). The second is located 7q22.1 in the 13th intron of the DNA replication gene MCM7 and consists of three miRNAs, miR-106b (MIR106B), miR-25 (MIR25) and miR-93 (MIR93).

Because miR~106a~363 and miR~106b~25 contain miRNAs that are highly similar, and in some cases identical, to those encoded by miR~17~92, it has been suggested that these clusters may have similar functions [Ventura et al. 2008].

Various cancer types (colon, pancreas, lung, gastric, leukemia) have been associated with alterations in the expression levels of miR~106a~363 cluster members, notably miR-106a. Members of the miR~106b~25 cluster accumulate in different types of cancer, including gastric, prostate, and pancreatic neuroendocrine tumors, neuroblastoma, and multiple myeloma [Ventura et al. 2008]. However, current literature has not demonstrated a significant variation of these miRNAs in breast cancer [Volinia et al. 2006].

A male/female breast cancer comparison showed a different expression of 17 miRNAs between the two categories, with 4 upregulated and 13 down-regulated (including miR-106a) miRNAs in male breast cancers [Fassan et al. 2009].

MIR-122

miR-122 has been described as up-regulated in breast cancer, compared to normal tissue [Iorio *et al.* 2005; Volinia *et al.* 2006].

MIR-124

This miRNA is a putative tumor suppressor [Lujambio and Esteller 2007]. Aberrant hypermethylation was shown mir-124-3 in a series of 71 primary human breast cancer specimens [Lehmann *et al.* 2008].

EZH2 is a multifunctional protein that integrates Wnt and E_2 signaling in BCC and the abnormal function of this protein is linked to several diseases (Shi *et al.* 2007). Based on TargetScan analysis, EZH2 is a likely target of mir-124a and mir-506, both of which are repressed by E_2 (in [Bhat-Nakshatri *et al.* 2009]).

MIR-125

miR-125b was one of the most consistently down-regulated miRNAs in breast cancer compared to normal tissue [Iorio *et al.* 2005]. Similarly, miR-125b-1 and miR-125b-2 were shown to be down-regulated in breast tumors, as compared to normal tissues [Volinia *et al.* 2006].

The ERBB family is a group of signaling proteins often deregulated in cancer and ERBB2 (better known as HER2/neu) gene amplification and protein over-expression are associated with an adverse outcome in breast cancer [Lacroix *et al.* 2004]. A bioinformatics search identified targeting seed sequences for miR-125a and miR-125b within the 3'-UTR of both HER2/neu and ERBB3. Using the human BCC line SK-BR-3 as a model for HER2/neu and ERBB3 dependence [Lacroix and Leclercq 2004a], infection of these cells with retroviral constructs expressing either miR-125a or miR-125b resulted in suppression of HER2/neu and ERBB3 at both the transcript and protein level. This effect was only marginal in the HER2/neu-independent BCC line MCF10A [Scott *et al.* 2007].

Human vitamin D3 hydroxylase (CYP24) catalyzes the inactivation of 1-alpha, 25-dihydroxyvitamin D3 (calcitriol), which exerts antiproliferative effects. *In silico* analysis identified a potential miR-125b recognition element

(MRE125b) in the 3'-UTR of human CYP24 mRNA. Immunohistochemical analysis revealed that the CYP24 protein levels in human breast cancer were higher than in adjacent normal tissues, without an accompanying CYP24 mRNA increase. On the other hand, the expression levels of miR-125b in cancer tissues were significantly lower than those in normal tissues. Interestingly, the CYP24 protein levels in cancer tissues were inversely associated with the cancer/normal ratios of the miR-125b levels, indicating that the decreased miR-125b levels in breast cancer tissues would be one of the causes of the high CYP24 protein expression [Komagata *et al.* 2009].

It was recently shown that a germline mutation in mature miR-125a was highly associated with breast cancer tumorigenesis, further supporting a role of miR-125a as a tumor suppressor gene in human cancer [Li *et al.* 2009a].

A male/female breast cancer comparison showed a different expression of 17 miRNAs between the two categories, with 4 upregulated and 13 down-regulated (including miR-125b and miR-125a-5p) miRNAs in male breast cancers [Fassan *et al.* 2009].

HuR is a stress-induced RNA binding protein whose expression is elevated or localization perturbed in several different cancers. Increased cytoplasmic localization of HuR is a prognostic marker in breast cancer. The expression of miR-125a was inversely correlated with HuR expression in several different BCC lines. Real time PCR and gene reporter assays indicated that HuR was translationally repressed by miR-125a. Re-establishing miR-125a expression in BCC decreased HuR protein level and inhibited cell growth. In MCF-7 BCC, it was shown that miR-125a inhibited cell growth via a dramatic suppression of cell proliferation and promotion of apoptosis. In addition, cell migration was also inhibited by miR-125a overexpression. Importantly, the repression of cell proliferation and migration engendered by miR-125a was partly rescued by HuR re-expression. All this suggests that miR-125a may function as a tumor suppressor for breast cancer, with HuR as a direct and functional target [Guo *et al.* 2009].

MIR-126

miR-126 was found down-regulated in BCC. Flow cytometry analysis showed that miR-126 inhibited cell cycle progression from G_1/G_0 to S. Further studies revealed that mir-126 targeted insulin receptor substrate (IRS)-1 at the translation level. IRS-1 is an adaptor protein in the insulin-like growth factor I (IGF-I)/IGF-I receptor (IGF-IR) pathway that mediate cell proliferation, migration, and survival. In breast cancer, IRS-1 is known as a transforming oncogene, being mainly involved in cell proliferation and survival [Chan and Lee 2008]. In accordance with this, knocking down of IRS-1 suppressed cell growth in HEK293 human embryonic kidney cells and in MCF-7 BCC, which recapitulated the effects of mir-126 [Zhang *et al.* 2008].

A search for general regulators of cancer metastasis has yielded a set of miRNAs for which expression was specifically lost in BCC developing metastatic potential. Restoring the expression of these miRNAs in malignant cells suppressed lung and bone metastasis by human cancer cells in vivo. Of these miRNAs, miR-126 restoration reduced overall tumor growth and proliferation. Expression of miR-126 was lost in the majority of primary breast tumors from patients who relapse, and its loss of expression of either miRNA was associated with poor distal metastasis-free survival. miR-126 was thus identified as metastasis suppressor miRNA in human breast cancer [Tavazoie *et al.* 2008].

In addition to breast cancer, miR-126 is down-regulated in various other tumor types, including hepatocellular carcinoma [Wong *et al.* 2008], cervical cancer [Wang *et al.* 2008b], colon cancer [Guo *et al.* 2008] and lung cancer [Liu *et al.* 2009]. Its expression suppresses the growth and/or invasion of cervical [Wang *et al.* 2008b], lung [Crawford *et al.* 2008; Liu et *al.* 2009], and colon [Guo *et al.* 2008] cancer cells. miR-126 thus seems to be a "general" tumor suppressor.

MIR-127

In an analysis involving 435 miRNA oligonucleotide probes, miR-127 was observed to be down-regulated in breast cancer compared to normal adjacent tumor tissues [Yan *et al.* 2008].

MIR-128

miR-128 expression has been associated to ER-positive/LN-negative breast cancer aggressiveness. Bioinformatics analysis coupled miR-128 to cytokine signaling [Foekens *et al.* 2008].

MIR-130

miR-130a has been shown to be down-regulated in breast tumors compared to normal tissues [Volinia et al. 2006]. Among breast tumors, miR-130 expression was relatively higher in ER-positive and in low-grade lesions [Blenkiron et al. 2007].

MIR-135B

In BCC, miR-135b was part of a miRNA signature corresponding with ER status identified by stepwise artificial neural networks analysis [Lowery et al. 2009]

MIR-140

miR-140 has been shown to be down-regulated in breast tumors compared to normal tissues [Volinia et al. 2006].

MIR-141

See miR-200 family

MIR-143

E_2 was found to repress seven miRNAs (miR-302b, miR-506, miR-524, miR-27a, miR-27b, miR-143, miR-9) in MCF-7 BCC [Bhat-Nakshatri et al. 2009].

MIR-145

miR-145 was one of the most consistently down-regulated miRNAs in breast cancer compared to normal tissue. Among breast tumors, miR-145 level was higher in lesions with a low proliferation index [Iorio et al. 2005].

Down-regulation of miR-145 in breast tumors has been repeatedly reported [Volinia *et al.* 2006; Sempere *et al.* 2008]. Sempere *et al.* (2008) implemented an *in situ* hybridization method to reveal the spatial distribution of miRNA expression in archived formalin-fixed, paraffin-embedded specimens representing normal and tumor tissue from >100 breast cancer cases. They observed that expression of miR-145 was restricted to the myoepithelial/basal cell compartment of normal mammary ducts and lobules, whereas its accumulation was reduced or completely eliminated in matching tumor specimens. These authors also observed early manifestation of altered miR-145 expression in atypical hyperplasia and carcinoma in situ lesions, suggesting that this miRNA may have a potential clinical application as a novel biomarker for early detection [Sempere *et al.* 2008].

In addition to breast cancer, significant miR-145 down-regulation has been shown in ovarian, prostate, colorectal, liver and ACTH-secreting pituitary cancers, supporting a general function of this miRNA as tumor suppressor [Iorio *et al.* 2007; Ozen *et al.* 2008; Slaby *et al.* 2007; Varnholt *et al.* 2008; Schepeler *et al.* 2008; Amaral *et al.* 2009].

The tumor suppressor p53 negatively regulates a number of genes, including the proto-oncogene MYC, in addition to activating many other genes. p53 was seen to transcriptionally induce the expression of miR-145 by interacting with a potential p53 response element (p53RE) in the miR-145 promoter. It was further shown that MYC was a direct target for miR-145. The blockade of miR-145 by anti-miR-145 was able to reverse the p53-mediated MYC repression. The specific silencing of MYC by miR-145 accounted at least in part for the miR-145-mediated inhibition of tumor cell growth both *in vitro* and *in vivo* [Sachdeva *et al.* 2009].

Overexpression of miR-145 by plasmid inhibited MCF-7 BCC growth and induced apoptosis. Subsequently, the Rho effector rhotekin (RTKN), involved in cancer cell survival, was identified as a potential miR-145 target by bioinformatics. Using reporter constructs, it was found that the RTKN 3' UTR carried the directly binding site of miR-145. Additionally, overexpression of miR-145 in MCF-7 BCC reduced RTKN protein expression as well as mRNA level. Furthermore, down-regulation of RTKN by siRNA was shown to inhibit MCF-7 cell growth [Wang *et al.* 2009c].

The effects of miR-145 re-expression in BCC lines were assayed by using proliferation and apoptosis assays. It was found that miR-145 exhibited a pro-apoptotic effect, which was dependent on TP53 activation, and that TP53 activation could, in turn, stimulate miR-145 expression, thus establishing a death-promoting loop between miR-145 and TP53. It was also found that miR-

145 could downregulate ER-α protein expression through direct interaction with two complementary sites within its coding sequence. These findings support a view that miR-145 re-expression therapy could be mainly envisioned in the specific group of patients with ER-α-positive and/or TP53 wild-type tumors [Spizzo et al. 2009].

MIR-146

miR-146 has been shown to be down-regulated in various tumor types (breast, pancreas, prostate) compared to normal tissues [Volinia et al. 2006].

miR-146 has been associated with reduced metastatic ability of breast cancer. Expression of miR-146a/b in the highly metastatic BCC line MDA-MB-231 was shown to impair NF-κB activity by downregulating interleukin (IL)-1 receptor-associated kinase and TNF receptor-associated factor 6, two key adaptor/scaffold proteins in the IL-1 and Toll-like receptor signaling pathway, known to positively regulate NF-κB activity. miR-146a/b-expressing MDA-MB-231 cells showed markedly impaired invasion and migration capacity relative to control cells [Bhaumik et al. 2008].

The protein BReast cancer Metastasis Suppressor 1 (BRMS1) was seen to significantly up-regulates miR-146a in MDA-MB-231 BCC and MDA-MB-435 melanoma [Lacroix 2009] cells, and miR-146b in MDA-MB-435. Transduction of miR-146a or miR-146b into MDA-MB-231 down-regulated expression of EGFR, inhibited invasion and migration *in vitro*, and suppressed experimental lung metastasis. These results suggest that modulating the levels of miR-146a or miR-146b could have a therapeutic potential to suppress breast cancer metastasis [Hurst et al. 2009].

A G to C polymorphism (rs2910164) is located within the sequence of miR-146a precursor, which leads to a change from a G:U pair to a C:U mismatch in its stem region. The predicted miR-146a target genes include BRCA1 and BRCA2, which are key breast and ovarian cancer genes. To examine whether rs2910164 plays any role in breast and/or ovarian cancer, associations between this polymorphism and age of diagnosis in patients with familial breast or ovarian cancer were studied. Patients who had at least one miR-146a variant allele were diagnosed at an earlier age than with no variant alleles. In further functional analysis, it was found that the variant allele displayed increased production of mature miR-146a from the precursor miRNA compared with the common allele. Consistent with the target prediction, in a target in vitro assay, it was observed that miR-146a could bind to the 3'-UTR of BRCA1 and BRCA2 messenger RNAs (mRNAs) and potentially modulate their mRNA expression [Shen *et al.* 2008].

MIR-148

Aberrant hypermethylation was shown for five miRNAs, including mir-148, in 34-86% of cases in a series of 71 primary human breast cancer specimens [Lehmann *et al.* 2008].

In a search of a miRNA hypermethylation profile characteristic of human metastasis, LN metastatic cancer cells were treated with a DNA demethylating agent followed by hybridization to an expression microarray. Among the miRNAs that were reactivated upon drug treatment, miR-9, miR-34b/c, and miR-148a were found to undergo specific hypermethylation-associated silencing in cancer cells compared with normal tissues. The reintroduction of miR-34b/c and miR-148a in cancer cells with epigenetic inactivation inhibited their motility, reduced tumor growth, and inhibited metastasis formation in xenograft models, with an associated down-regulation of the miRNA oncogenic target genes, such as MYC, E2F3, CDK6, and TGIF2. Most important, the involvement of miR-9, miR-34b/c, and miR-148a hypermethylation in metastasis formation was also suggested in human primary malignancies (n = 207) because it was significantly associated with the appearance of LN metastasis. This shows that DNA methylation-associated silencing of tumor suppressor miRNAs, including miR-148 might contribute to the development of human cancer metastasis [Lujambio *et al.* 2008].

MIR-152

Aberrant hypermethylation was shown for mir-152 in a series of 71 primary human breast cancer specimens [Lehmann *et al.* 2008].

MIR-155

miR-155 was described as one of the most consistently up-regulated miRNAs in breast cancer compared to normal tissue [Iorio *et al.* 2005; Volinia *et al.* 2006, Yan *et al.* 2008]. miR-155 is viewed as an oncogene. In fact, high miR-155 expression has been associated to cancer and cell proliferation in a number of reports. This miRNA is notably overexpressed in lymphoma (reviewed in [Turner and Vigorito 2008]), chronic lymphocytic leukemia [Fulci *et al.* 2007; Wang *et al.* 2008], lung [Yanaihara *et al.* 2006], pancreas [Lee *et al.* 2007; Habbe *et al.* 2009], thyroid [Nikiforova *et al.* 2008], cervical [Wang *et al.* 2008], and clear cell renal [Jung *et al.* 2009] cancers. miR-155 appears as a typical multifunctional miRNA, being involved not only in cancer, but also in various physiological and pathological processes such as hematopoietic lineage differentiation, immunity, inflammation, and cardiovascular diseases (reviewed in [Faraoni *et al.* 2009]).

By hybridizing a 515-miRNA oligonucleotide-based microarray library, a total of 28 miRNAs were found to be significantly deregulated in TGF-β-treated normal murine mammary gland (NMuMG) epithelial cells but not SMAD4 knockdown NMuMG cells. Among up-regulated miRNAs, miR-155 was the most significantly elevated miRNA. TGF-β induced miR-155 expression and promoter activity through SMAD4. The knockdown of miR-155 suppressed TGF-β-induced epithelial-mesenchymal transition (EMT, see miR-200 family) and tight junction dissolution, as well as cell migration and invasion. Further, the ectopic expression of miR-155 reduced RhoA protein and disrupted tight junction formation. Reintroducing RhoA cDNA without the 3'-UTR largely reversed the phenotype induced by miR-155 and TGF-β. In addition, elevated levels of miR-155 were frequently detected in invasive breast cancer tissues. Thus, miR-155 may play an important role in TGF-β-induced EMT and cell migration and invasion by targeting RhoA and possess a potential therapeutic target for breast cancer intervention [Kong *et al.* 2008].

MIR-181

miR-181 miRNA was shown to be induced in cancer cell lines placed in hypoxic conditions [Kulshreshtha *et al.* 2007].

miR-181b-1 was found up-regulated in various tumor types (breast, pancreas, prostate), while miR-181a-1 was up-regulated only in breast tumors compared to normal tissues [Volinia *et al.* 2006]. miR-181b up-regulation in breast cancer compared to normal adjacent tumor tissues was found in an analysis involving 435 miRNA oligonucleotide probes [Yan *et al.* 2008].

A miRNA microarray analysis of MCF-7 BCC lines that are either sensitive (parental) or resistant to tamoxifen showed marked up-regulation of miR-221, miR-222 and miR-181 in resistant cells compared with parental

MCF-7 cells [Miller et al. 2008]. miR-221 and miR-222 have been shown to repress p27 expression in various cancer cells, and this repression promotes tumor cell proliferation. Of interest, a study in HL-60 myeloid cells has shown that p27 mRNA is translated via a cap-dependent mechanism in and that this translation is repressed by miR-181a [Cuesta et al. 2009].

In BCC, miR-181c was part of a miRNA signature corresponding with HER2/neu receptor status identified by stepwise artificial neural networks analysis [Lowery et al. 2009].

In BCC, the expression of a broad set of miRNAs, including miR-181a, miR-181b and miR-181c, was decreased following E_2 treatment in an ER-dependent manner. In addition, a transcriptome analysis revealed that the E_2-repressed miR-26a and miR-181a regulate many genes associated with cell growth and proliferation, including the progesterone receptor gene, a key actor in estrogen signaling. Strikingly, miRNA expression is also regulated in breast cancers of women who had received antiestrogen neoadjuvant therapy [Maillot et al. 2009].

A male/female breast cancer comparison showed a different expression of 17 miRNAs between the two categories, with 4 upregulated and 13 down-regulated (including miR-181c) miRNAs in male breast cancers [Fassan et al. 2009].

MIR~183~96~182 CLUSTER

Located at 7q32.2, this cluster is comprised of miR-96, miR-182 and miR-183.

Human breast tumors contain a breast cancer stem cell (BCSC) population with properties reminiscent of normal stem cells. A study found 37 microRNAs that were differentially expressed between human BCSCs and nontumorigenic cancer cells. Three miRNA clusters, miR~200c~141, miR~200b~200a~429, and miR~183~96~182 were downregulated in human BCSCs, but also in normal human and murine mammary stem/progenitor cells [Shimono et al. 2009].

The FOXO1 transcription factor orchestrates the regulation of genes involved in the apoptotic response, cell cycle checkpoints, and cellular metabolism. FOXO1 is a putative tumor suppressor. FOXO1 mRNA was shown to be down-regulated in breast tumor samples as compared with normal breast tissue. Silencing of the miRNA processing enzymes, Drosha and Dicer, led to an increase in FOXO1 expression. Functional and specific miRNA target sites in the FOXO1 3'-UTR for miR-27a, miR-96, and miR-182 were identified; these miRNAs have previously been linked to oncogenic transformation. In MCF-7 BCC, in which the level of FOXO1 protein is very low, all three miRNAs were observed to be highly expressed. Antisense inhibitors to each of these miRNAs led to a significant increase in endogenous FOXO1 expression and to a decrease in cell number in a manner that was blocked by FOXO1 siRNA. Overexpression of FOXO1 resulted in decreased cell viability because of inhibition of cell cycle traverse and induction of cell death. Thus, FOXO1 regulation, and targeting of FOXO1 by miRNAs may contribute to transformation or maintenance of an oncogenic state in BCC [Guttilla and White 2009].

MIR-190

In BCC, miR-190 was part of a miRNA signature corresponding with ER status identified by stepwise artificial neural networks analysis [Lowery *et al.* 2009]

MIR-193

miR-193b is one of five potent ER-α-regulating miRNAs, the other being miR-18a, miR-18b, miR-206 and miR-302c, confirmed to directly target ER-α in 3'-UTR reporter assays [Leivonen *et al.* 2009]. On the other hand, in MCF-7 BCC, the expression of a broad set of miRNAs, including miR-193b, was decreased following E_2 treatment in an ER-dependent manner [Maillot *et al.* 2009].

A study using miRNA arrays revealed that miR-193b was identified as differentially expressed between the MDA-MB-231 BCC line and its highly metastatic variant. A bioinformatics search revealed a potential target site for miR-193b within the 3'UTR of urokinase-type plasminogen activator (uPA). Ectopic expression of miR-193b repressed the expression of sensor constructs harboring the 3'UTR of uPA in BCC lines. Anti-miR-193b treatment led to an increase of uPA protein and increased cell invasion in MDA-MB-231 cells. In contrast, overexpression of miR-193b significantly reduced uPA protein amounts and inhibited cell invasion in MDA-MB-231 BCC as well as in MDA-MB-435 melanoma cells. In an immunodeficient mouse model, miR-

193b significantly inhibited the growth and dissemination of xenograft tumors. Immunohistochemical staining and real-time PCR assays showed that miR-193b was a negative regulator of the uPA gene in primary breast tumors [Li et al. 2009b].

MIR-196

Suppression of annexin A1 (ANXA1), a mediator of apoptosis and inhibitor of cell proliferation, is well documented in various cancers, including breast tumors [Ou et al. 2008; Chuthapisith et al. 2009] but the underlying mechanism remains unknown. miRNA-196a showed significant inverse correlation with ANXA1 mRNA levels in 12 cancer cell lines of esophageal, breast and endometrial origin, identifying it as the candidate miRNA targeting ANXA1. Inverse correlation was also observed in 10 esophageal adenocarcinomas. Analysis of paired normal/tumor tissues from additional 10 patients revealed an increase in miR-196a in the cancers, accompanied by a decrease in ANXA1 mRNA. Increasing miR-196a levels in cells by miR-196a mimics resulted in decreased ANXA1 mRNA and protein. In addition, miR-196a mimics inhibited luciferase expression in luciferase plasmid reporter that included predicted miR-196a recognition sequence from ANXA1 3'-UTR confirming that miR-196a directly targets ANXA1. miR-196a promoted cell proliferation, anchorage-independent growth and suppressed apoptosis, suggesting its oncogenic potential [Luthra et al. 2008].

In a case-control study of 1,009 breast cancer cases and 1,093 cancer-free controls in a population, the mir-196a-2 rs11614913:T>C single nucleotide polymorphism (SNP) was associated with significantly increased risks of breast cancer in a dose-effect manner. This suggests that SNP in miRNAs may contribute to breast cancer susceptibility [Hu et al. 2009].

A genetic association analysis was performed by screening genetic variants in 15 miRNA genes. This analysis detected that a common sequence variant in miR-196a-2 (rs11614913, C-->T) was significantly associated with decreased breast cancer risk. Hypermethylation of a CpG island upstream (-700 bp) of the miR-196a-2 precursor was also associated with reduced breast cancer risk. By delivering expression vectors containing either wild-type or mutant precursors of miR-196a-2 into BCC, we showed that this variant led to less efficient processing of the miRNA precursor to its mature form as well as diminished capacity to regulate target genes. A whole-genome expression microarray was done and a pathway-based analysis identified a cancer-relevant network formed by genes significantly altered following enforced expression of miR-196a-2. Mutagenesis analysis further showed that cell cycle response to mutagen challenge was significantly enhanced in cells treated with variant miR-196a-2 compared with cells treated with the wild-type. Thus, miR-196a-2 might have a potentially oncogenic role in breast tumorigenesis, and the functional genetic variant in its mature region could serve as a novel biomarker for breast cancer susceptibility [Hoffman *et al.* 2009].

MIR-199

A male/female breast cancer comparison showed a different expression of 17 miRNAs between the two categories, with 4 upregulated and 13 down-regulated (including miR-199a-5p) miRNAs in male breast cancers [Fassan *et al.* 2009].

MIR-200 FAMILY (MIR~200B~200A~429 AND MIR~200C~141 CLUSTERS)

This family is comprised of five members: miR-141, miR-200a, miR-200b, miR-200c, miR-429, corresponding to two clusters, miR~200b~200a~429 (at 1p36.33) and miR~200c~141 (at 12p13.31).

Various studies have shown that expression of this family is associated with the phenotype of BCC and with epithelial-mesenchymal transition (EMT). Most BCC have an "epithelial-like" phenotype, notably characterized by a high level of E-cadherin and steroid receptors expression (as observed, for instance, in the MCF-7 BCC line); other BCC may rather have a mesenchymal-like phenotype, notably characterized by the absence or a low level of E-cadherin (as exemplified by the MDA-MB-231 BCC line) (for a review, see [Lacroix and Leclercq 2004a]). The epithelial-mesenchymal transition (frequently shortened to EMT) hypothesis postulates that BCC are able to convert their phenotype and that metastasis could need a transitory EMT event. Indeed, it has been observed that EMT may be induced in BCC by factors such as TGF-β or bone morphogenetic proteins (BMPs), notably acting through up-regulation of the E-cadherin transcriptional repressors ZEB1 (also known as δEF1 or transcription factor 8) and ZEB2 (also known as SIP1 or ZFHX1B).

An increasing series of articles support that a high expression level of miR-200 family members is positively correlated to the epithelial phenotype and E-cadherin expression, while being inversely correlated to ZEB1 and ZEB2 expression [Hurteau *et al.* 2007; Sempere *et al.* 2007; Gregory *et al.* 2008; Burk *et al.* 2008; Bracken *et al.* 2008, Cochrane *et al.* 2009]. Moreover, it appears that these miRNAs, the expression of which is absent or lost in invasive BCC lines with mesenchymal phenotype and in regions of metaplastic breast cancer specimens lacking E-cadherin, target ZEB1 and ZEB2, as well as the EMT inducer TGF-β2. They strongly activate epithelial differentiation in pancreatic, colorectal and BCC. Their enforced expression is sufficient to prevent TGF-β-induced EMT. It has been suggested that loss of expression of the miR-200 family members during cancer progression may play a critical role in the repression of E-cadherin by ZEB1 and ZEB2 during EMT, thereby enhancing migration and invasion [Korpal *et al.* 2008; Korpal and Kang 2008].

The situation is complicated by the fact that, in turn, ZEB1 and ZEB2 are able to repress the expression of all miR-200 family members [Burk *et al.* 2008; Bracken *et al.* 2008]. For instance, miRNA-200a, miR-200b, and miR-429 are all encoded on a 7.5-kb polycistronic primary miRNA (pri-miR) transcript. The promoter for the pri-miR is located within a 300-bp segment located 4 kb upstream of miR-200b and is sufficient to confer expression in epithelial cells, while it is repressed in mesenchymal cells by ZEB1 and ZEB2 through their binding to a conserved pair of ZEB-type E-box elements located proximal to the transcription start site. ZEB1 has also been shown to suppress transcription of miR-141 and miR-200c. Thus, a double-negative feedback loop exists between ZEB1-ZEB2 and the miRNA-200 family and appears to play an essential role in regulating EMT in BCC.

Human breast tumors contain a breast cancer stem cell (BCSC) population with properties reminiscent of normal stem cells. A study found 37 miRNAs that were differentially expressed between human BCSCs and nontumorigenic cancer cells. Three clusters, miR~200c~141, miR~200b~200a~429, and miR~183~96~182 were downregulated in human BCSCs, normal human and murine mammary stem/progenitor cells, and embryonal carcinoma cells. Expression of BMI1, a known regulator of stem cell self-renewal, was modulated by miR-200c. miR-200c inhibited the clonal expansion of BCC and suppressed the growth of embryonal carcinoma cells in vitro. Most importantly, miR-200c strongly suppressed the ability of normal mammary stem cells to form mammary ducts and tumor formation driven by human BCSCs in vivo. The coordinated downregulation of three miRNA clusters and the similar functional regulation of clonal expansion by miR-200c provide a molecular link that connects BCSCs with normal stem cells [Shimono *et al.* 2009].

miR-200a, miR-200b and miR-200c were shown to be induced by E_2 in BCC [Bhat-Nakshatri *et al.* 2009]. On the other hand, in MCF-7 BCC, the expression of a broad set of miRNAs, including miR-200a and miR-200c, was decreased following E_2 treatment, in an ER-dependent manner [Maillot *et al.* 2009].

In BCC with low or absent miR-200c, reinstatement of miRNA-200c was shown to restore E-cadherin and dramatically reduce migration and invasion. Microarray profiling revealed that in addition to ZEB1 and ZEB2, other mesenchymal genes (such as FN1, NTRK2, and QKI), which are also predicted direct targets of miR-200c, were indeed inhibited by addition of exogenous miR-200c. One such gene, class III beta-tubulin (TUBB3), which encodes a tubulin isotype normally found only in neuronal cells, is a direct target of miR-200c. Restoration of miR-200c increased sensitivity of cancer cells to microtubule-targeting agents. Because expression of TUBB3 is a common mechanism of resistance to microtubule-binding chemotherapeutic agents in many types of solid tumors, the ability of miR-200c to restore chemosensitivity to such agents may be explained by its ability to reduce TUBB3 [Cochrane *et al.* 2009].

Akt family of serine/threonine-specific protein kinases is composed of three isoforms: Akt1, Akt2 and Akt3. Although Akt is known to play a role in human cancer, the relative contribution of its three isoforms to oncogenesis remains to be determined. It was shown that the expression of miR-200 family members was correlated to the Akt1/Akt2 ratio. The ratio of Akt1 to Akt2 and the abundance of miR-200 and of the messenger RNA encoding E-cadherin in a set of primary and metastatic human breast cancers were consistent with the hypothesis that in many cases breast cancer metastasis may be under the control of the Akt-miR-200-E-cadherin axis. Thus, it seems that induction of EMT is controlled by miRNAs whose abundance depends on the balance between Akt1 and Akt2 rather than on the overall activity of Akt [Iliopoulos *et al.* 2009].

A male/female breast cancer comparison showed a different expression of 17 miRNAs between the two categories, with 4 upregulated and 13 down-regulated (including miR-200b) miRNAs in male breast cancers [Fassan *et al.* 2009].

MIR-203

miR-203 was induced by E_2 in MCF-7/p BCC [Bhat-Nakshatri et al. 2009]. However, in MCF-7 cells, the expression of a broad set of miRNAs, including miR-203, was decreased following E_2 treatment in an ER-dependent manner [Maillot et al. 2009].

MIR-205

miR-205 has been shown to be down-regulated in breast tumors compared to normal tissues [Volinia et al. 2006]. Among breast tumors, miR-205 expression was associated with absence of vascular invasion [Iorio et al. 2005].

As also observed for miR-200 family members, to which it is functionally related, miR-205 was markedly down-regulated in cells that had undergone EMT in response to TGF-β or to ectopic expression of the protein tyrosine phosphatase Pez [Gregory et al. 2008].

Another study confirmed the underexpression of miR-205 in breast cancer compared to the matched normal breast tissue. Moreover, BCC lines, including the widely used MCF-7 and MDA-MB-231 were shown to express a lower level of miR-205 than the non-malignant MCF-10A cells. Ectopic expression of miR-205 significantly inhibited cell proliferation and anchorage independent growth, as well as cell invasion. Furthermore, miR-205 was shown to suppress lung metastasis in an animal model. Western blot combined with the luciferase reporter assays demonstrated that ERBB3 and vascular endothelial growth factor A (VEGF-A) were direct targets for miR-205. Together, these results suggest that miR-205 is a tumor suppressor in breast cancer [Wu et al. 2009; Wu and Mo 2009].

HER2/neu overexpression is a hallmark of a particularly aggressive subset of breast tumors, and its activation is strictly dependent on the trans-interaction with other members of ERBB family; in particular, the activation of the PI3K/Akt survival pathway, so critically important in tumorigenesis, is predominantly driven through phosphorylation of the kinase-inactive member ERBB3. miR-205 was shown to directly target ERBB3 and to inhibit the activation of the downstream mediator Akt. miR-205 reintroduction in SK-BR-3 cells inhibited their clonogenic potential and increased the responsiveness to tyrosine-kinase inhibitors gefitinib and lapatinib, abrogating the ERBB3-mediated resistance and restoring a potent proapoptotic activity. miR-205 is thus able to interfere with the proliferative pathway mediated by ERBB receptor family, and the data suggest that miR-205 could improve the responsiveness to specific anticancer therapies [Iorio *et al.* 2009].

An *in situ* hybridization method was implemented to reveal the spatial distribution of miRNA expression in archived formalin-fixed, paraffin-embedded specimens representing normal and tumor tissue from >100 breast cancer cases. Expression of miR-205 was shown to be restricted to the myoepithelial/basal cell compartment of normal mammary ducts and lobules, whereas its accumulation was reduced or completely eliminated in matching tumor specimens [Sempere *et al.* 2008].

MIR-206

miR-206 was found up-regulated in breast cancer compared to normal tissue. Its expression level was relatively higher in ER-negative breast cancer [Iorio *et al.* 2005]. This is in agreement with the fact that miR-206 may down-regulate the expression of ER-α in breast cancer. Two putative miR-206 sites were found *in silico* within the 3'-UTR of human ER-α mRNA. Transfection of MCF-7 BCC with pre-miR-206 or 2'-O-methyl antagomiR-206 specifically decreased or increased, respectively, ER-α mRNA levels [Adams *et al.* 2007]. The ability of miR-206 to potently regulate ER-α expression by directly targeting the 3'-UTR of ER-α mRNA was confirmed by reporter assays [Leivonen *et al.* 2009]. miR-206 expression was markedly decreased in ER-α-positive human breast cancer tissues assayed by quantitative reverse transcription-PCR analysis. In agreement with this, miR-206 expression was higher in ER-α-negative MDA-MB-231 than ER-α-positive MCF-7 BCC [Adams *et al.* 2007; Kondo *et al.* 2008]. miR-206 expression was strongly inhibited by ER-α agonists, indicating a mutually inhibitory feedback loop [Adams *et al.* 2007].

EGFR/MAPK signaling can induce a switch in MCF-7 BCC, from an ER-α-positive, luminal-A phenotype, to an ER-α-negative, basal-like phenotype. Although mechanisms for this switch remain obscure, basal-like cancers are typically high grade and confer a poorer clinical prognosis [Lacroix *et al.* 2004]. It was previously reported (see [Adams *et al.* 2007] here above) that miR-206 and ER-α could repress each other's expression in MCF-7 BCC in a double-negative feedback loop. miR-206 was further shown to coordinately target mRNAs encoding the steroid receptor coactivator proteins SRC-1 and

SRC-3 (also known as NCOA3, see miR-17), and the transcription factor GATA-3 [Lacroix and Leclercq 2004b], all of which contribute to estrogenic signaling and a luminal-A phenotype. Overexpression of miR-206 repressed estrogen-mediated responses in MCF-7 BCC, even in the presence of ER-α encoded by an mRNA lacking a 3'-UTR, suggesting miR-206 affects estrogen signaling by targeting mRNAs encoding ER-α-associated coregulatory proteins. Furthermore, EGF treatments enhanced miR-206 levels in MCF-7 cells and ER-α-negative, EGFR-positive MDA-MB-231 BCC, whereas EGFR small interfering RNA, or PD153035, an EGFR inhibitor, or U0126, a MAPK kinase inhibitor, significantly reduced miR-206 levels in MDA-MB-231 cells. Blocking EGF-induced enhancement of miR-206 with antagomiR-206 abrogated the EGF-inhibitory effect on ER-α, SRC-1, and SRC-3 levels, and on estrogen response element-luciferase activity, indicating that EGFR signaling represses estrogenic responses in MCF-7 cells by enhancing miR-206 activity. Elevated miR-206 levels in MCF-7 BCC ultimately resulted in reduced cell proliferation, enhanced apoptosis, and reduced expression of multiple estrogen-responsive genes. In conclusion, miR-206 contributes to EGFR-mediated abrogation of estrogenic responses in MCF-7 cells, contributes to a luminal-A- to basal-like phenotypic switch, and may be a measure of EGFR response within basal-like breast tumors [Adams et al. 2009].

miR-206 was found down-regulated in highly metastatic variants of the MDA-MB-231 BCC. Its ectopic expression in the metastatic variants reduced both lung and bone metastases when such cells were injected directly into blood circulation. When MDA-MB-231 BCC were injected into the mammary fat pads of immunodeficient mice, restoring the expression of miR-206 did not affect either the proliferation or apoptosis of primary tumor cells but did alter cell morphology and inhibited motility and invasiveness. Hence, miRNAs can suppress metastasis through their ability to inhibit cell migration, invasion or proliferation [Tavazoie et al. 2008].

MiR-210

miR-210 was found up-regulated in breast and lung tumors compared to normal tissues [Volinia et al. 2006].

miR-210 has been associated to ER-positive/LN-negative breast cancer aggressiveness. It has also been associated to metastatic capability in ER-negative/LN-negative breast cancer and in the clinically important triple-negative (ER-negative, PR-negative, HER2/neu-negative) tumor subgroup.

Several recent studies have established a link between hypoxia, a well-documented component of the tumor microenvironment, and specific miRNAs. One member of this class is miR-210 [Kulshreshtha et al. 2007; Wang and Lee 2009], which was identified as hypoxia-inducible in all cell types tested, and was shown to be overexpressed in most cancer types [Ivan et al. 2008; Giannakakis et al. 2008; Crosby et al. 2009]. Bioinformatic analysis coupled miR-210 to hypoxia/VEGF signaling [Foekens et al. 2008].

In breast cancer, miR-210 overexpression was induced by hypoxia in a hypoxia-inducible factor (HIF)-1α- and von Hippel-Lindau (VHL) protein-dependent fashion and its expression level in breast cancer samples was found to constitute an independent prognostic factor [Camps et al. 2008].

One important target of miR-210 is MNT, a known MYC antagonist. MNT mRNA contains multiple miR-210 binding sites in its 3' UTR. This suggests that miR-210 may influence the hypoxia response in tumor cells through targeting a key transcriptional repressor of the MYC-MAX network [Zhang et al. 2009].

Biocomputational analysis and *in vitro* assays demonstrated that E2F transcription factor 3 (E2F3), a key protein in cell cycle, was regulated by miR-210. E2F3 was further confirmed to be down-regulated at the protein level upon induction of miR-210 [Giannakakis et al. 2008].

Forced expression of miR-210 was found to suppress the levels of RAD52, which is a key factor in homology-dependent repair (HDR) and miR-210 was capable of interacting with the 3' UTR of RAD52 [Crosby et al. 2009].

MIR-213 = MIR-181A-1

miR-213 was found up-regulated in breast and lung tumors, as compared to normal tissues [Iorio et al. 2005; Volinia et al. 2006]. Its expression level was higher in tumor stages 2-3 compared to stage 1 [Iorio et al. 2005].

A miRNA microarray analysis of MCF-7 BCC lines that are either sensitive or resistant to tamoxifen showed marked up-regulation of miR-213 in resistant cells compared with sensitive (parental) MCF-7 cells [Miller et al. 2008].

MiR-217

In BCC, miR-217 was part of a miRNA signature corresponding with ER-α status identified by stepwise artificial neural networks analysis [Lowery et al. 2009].

MiR-218

In BCC, miR-218 was part of a miRNA signature corresponding with ER-α status identified by stepwise artificial neural networks analysis [Lowery et al. 2009].

MiR~222~221 Cluster

It has been speculated that the miRNAs comprised in a same cluster might play related biological functions. It is now clear that miRNAs in two clusters (miR-106b~93~25 and miR~222~221) suppress the Cip/Kip family members of cyclin-dependent kinase (CDK) inhibitors p57 (or Kip2), p21 (or Cip1) and p27 (or Kip1) [Medina et al. 2008; Kim et al. 2009]. Indeed, the effects of miR-221 and miR-222 on p27 have been widely described. miR-221 and miR-222 repress p27 expression in various cancer cells, and this repression promotes tumor cell proliferation [Le Sage et al. 2007; Visone et al. 2007; Mercatelli et al. 2008; Medina et al. 2008, Cuesta et al. 2009].

miR-221 and miR-222, encoded in tandem Xp11.3, are coexpressed in almost all cancer types analyzed as yet, and their overexpression has been repeatedly associated with increased proliferation [He et al. 2005; Pallante et al. 2006; Lee et al. 2007; Galardi et al. 2007; Le Sage et al. 2007; Visone et al. 2007; Nikiforova et al. 2008; Chen et al. 2008; Zhang et al. 2008; Mercatelli et al. 2008]. It has also been shown that miR-221/miR-222 were able to down-regulate stem cell factor receptor c-KIT mRNA and protein [He et al. 2005; Poliseno et al. 2006; Felicetti et al. 2008; Kuehbacher et al. 2008]. c-KIT has been seen as a potential target in breast cancer therapy.

mir-221/miR-222 expression appears higher in ER–α-negative (MDA-MB-231, BT-549, Hs578T) than in ER-α-positive (MCF-7, T-47D) BCC lines [Sempere et al. 2008]. A search for regulators of ER-α expression yielded a set of miRNAs for which expression was specifically elevated in ER-α-negative breast cancer. Of these, miR-221 and miR-222 were shown to directly interact with the 3'-UTR of ER-α. Ectopic expression of miR-221 and miR-222 in MCF-7 and T-47D cells resulted in a decrease in expression of ER-α protein but not mRNA, whereas knockdown of miR-221 and miR-222 partially restored ER-α in ER-α protein-negative/mRNA-positive cells. Notably, miR-221- and/or miR-222-transfected MCF-7 and T-47D cells became resistant to the antiestrogen tamoxifen compared with vector-treated cells. Furthermore, knockdown of miR-221 and/or miR-222 sensitized MDA-MB-468 cells to tamoxifen-induced cell growth arrest and apoptosis. These findings indicate that miR-221 and miR-222 play a significant role in the regulation of ER-α expression at the protein level and could be potential targets for restoring ER-α expression and responding to antiestrogen therapy in a subset of breast cancers [Zhao et al. 2008].

In a comparative analysis of 4-hydroxytamoxifen (OHT)-sensitive and OHT-resistant MCF-7 BCC, OHT-resistant cells showed significant miR-221 and miR-222 up-regulation. The expression of miR-221 and miR-222 was also significantly elevated in HER2/neu-positive primary human breast cancer tissues that are known to be resistant to endocrine therapy compared with HER2/neu-negative tissue samples. Ectopic expression of miR-221/222 rendered the parental MCF-7 cells resistant to tamoxifen. The protein level of the cell cycle inhibitor p27 was reduced by 50% in OHT-resistant and by 28-50% in miR-221/222-overexpressing MCF-7 BCC. Furthermore, overexpression of p27 in the OHT-resistant BCC caused enhanced cell death when exposed to tamoxifen. These observations support the application of altered expression of specific miRNAs as a predictive tamoxifen-resistant breast cancer marker [Miller et al. 2008].

MIR-223

A male/female breast cancer comparison showed a different expression of 17 miRNAs between the two categories, with 4 upregulated and 13 down-regulated (including miR-223) miRNAs in male breast cancers [Fassan et al. 2009].

MIR-224

miR-224 down-regulation was noted in breast tumors compared to normal tissue [Volinia et al. 2006]. Among tumors, higher miR-224 level was found in ER-α-positive lesions [Blenkiron et al. 2007].

MIR-299

In BCC, miR-299 was part of a miRNA signature corresponding with ER-α status identified by stepwise artificial neural networks analysis [Lowery et al. 2009].

MIR-302

E_2 was found to repress seven miRNAs (miR-302b, miR-506, miR-524, miR-27a, miR-27b, miR-143, miR-9) in MCF-7 BCC [Bhat-Nakshatri et al. 2009].

In BCC, miR-302c was part of a miRNA signature corresponding with ER-α status identified by stepwise artificial neural networks analysis [Lowery et al. 2009]. The ability of miR-302c to potently regulate ER-α expression by directly targeting the 3'-UTR region of ER-α mRNA was shown by reporter assays [Leivonen et al. 2009].

MIR-320

In an analysis involving 435 miRNA oligonucleotide probes, miR-320a was found down-regulated in breast cancer compared to normal adjacent tumor tissues [Yan et al. 2008].

MIR-326

Multidrug resistance-associated protein (MRP-1/ABCC1) transports a wide range of therapeutic agents and may play a critical role in the development of multidrug resistance (MDR) in tumor cells. However, the regulation of MRP-1 remains controversial. The VP-16-resistant MDR cell line, MCF-7/VP was found to overexpress MRP-1 mRNA and protein compared to its parent BCC line, MCF-7. miR-326 was downregulated in MCF-7/VP compared to MCF-7. Additionally, miR-326 was downregulated in a panel of advanced breast cancer tissues and consistent reversely with expression levels of MRP-1. Furthermore, the elevated levels of miR-326 in the mimics-transfected VP-16-resistant BCC line, MCF-7/VP, downregulated

MRP-1 expression and sensitized these cells to VP-16 and doxorubicin. This suggests that miR-326 may be an efficient agent for preventing and reversing MDR in tumor cells [Liang et al. 2009].

MIR-328

Breast cancer resistance protein (BCRP/ABCG2) is a molecular determinant of pharmacokinetic properties of many drugs in humans. miR-328 was found to readily target the 3'-UTR of ABCG2. An inverse relation between the levels of miR-328 and ABCG2 in MCF-7 BCC and its ABCG2-overexpressing, mitoxantrone-resistant derivative MCF-7/MX100 BCC line was observed. miR-328 levels could be rescued in MCF-7/MX100 BCC by transfection with miR-328 plasmid. Luciferase reporter assays showed that ABCG2 3'-UTR-luciferase activity was decreased more than 50% in MCF-7/MX100 BCC after transfection with miR-328 plasmid, the activity was increased over 100% in MCF-7 BCC transfected with a miR-328 antagomir, and disruption of miR-328 response element within ABCG2 3'-UTR led to a 3-fold increase in luciferase activity. Furthermore, the level of ABCG2 protein was down-regulated when miR-328 was over-expressed, and the level was up-regulated when miR-328 was inhibited by selective antagomir. Altered ABCG2 protein expression was associated with significantly declined or elevated levels of ABCG2 3'-UTR and coding sequence mRNAs, suggesting possible involvement of the mechanism of mRNA cleavage. Finally, miR-328-directed down-regulation of ABCG2 expression in MCF-7/MX100 BCC resulted in increased mitoxantrone sensitivity. This suggest that miR-328 targets ABCG2 3'-UTR and, consequently, controls ABCG2 protein expression and influences drug disposition in human BCC [Pan et al. 2009].

MIR-335

A search for general regulators of cancer metastasis yielded a set of miRNAs, including miR-335, for which expression was specifically lost as human BCC develop metastatic potential. Restoring the expression of this miRNA in MDA-MB-231 BCC suppressed lung and bone metastasis by these cells *in vivo*. In cell culture, overexpression of miR-335 resulted in a reduction of invasive capacity *in vitro*. The potential role of miR-335 was supported by examination of its expression within clinical breast tumor samples; low miR-335 expression levels were associated with very poor metastasis-free survival. miR-335 was seen to regulate a set of genes whose collective expression in a large cohort of human tumors is associated with risk of distal metastasis. Notably, miR-335 suppressed metastasis and migration through targeting of the progenitor cell transcription factor SOX4 and extracellular matrix component tenascin C (TNC). Expression of miR-335 was lost in the majority of primary breast tumors from patients who relapse, and its loss of expression was associated with poor distal metastasis-free survival. miR-335 was thus identified as metastasis suppressor miRNA in human breast cancer [Tavazoie *et al.* 2008].

MIR-342

A miRNA microarray analysis of MCF-7 BCC lines that are either sensitive (parental) or resistant to tamoxifen showed marked down-regulation of miR-342 miRNA in resistant cells compared with sensitive MCF-7 BCC [Miller *et al.* 2008]. Among breast tumors, miR-342 level was higher in luminal A subtype, in ER-α-positive, and in low-grade lesions [Blenkiron *et al.* 2007].

In BCC, miR-342 was part of a miRNA signature corresponding with ER status identified by stepwise artificial neural networks analysis. Its expression was further analyzed in 95 breast tumors. miR-342 level was highest in ER- and HER2/neu-positive luminal B tumors and lowest in triple-negative (ER-negative, PR-negative, HER2/neu-negative) tumors [Lowery *et al.* 2009].

MIR-365

miR-365 was found up-regulated in breast cancer compared to normal adjacent tumor tissues [Yan *et al.* 2008].

MIR-373

To identify potential metastasis-promoting miRNAs, a genetic screen was performed, using a non-metastatic BCC line that was transduced with a miRNA-expression library and subjected to a trans-well migration assay. It was found that miR-373 and miR-520c stiulated cancer cell migration and invasion *in vitro* and *in vivo*, and that certain cancer cell lines depended on endogenous miR-373 activity to migrate efficiently. Mechanistically, the migration phenotype of miR-373 and miR-520c could be explained by suppression of CD44. A significant up-regulation of miR-373 was found in clinical breast cancer metastasis samples that correlated inversely with CD44 expression. Taken together, these findings indicate that miRNAs are involved in tumor migration and invasion, and implicate miR-373 and miR-520c as metastasis-promoting miRNAs [Huang *et al.* 2008].

MIR-376

In BCC, miR-376b was part of a miRNA signature corresponding with HER2/neu receptor status identified by stepwise artificial neural networks analysis [Lowery *et al.* 2009]

MIR-377

In BCC, miR-377 was part of a miRNA signature corresponding with progesterone receptor status identified by stepwise artificial neural networks analysis [Lowery *et al.* 2009]

MIR-424

Two different teams found that miR-424 was induced by E_2 in BCC [Bhat-Nakshatri *et al.* 2009; Castellano *et al.* 2009].

MIR-425

A male/female breast cancer comparison showed a different expression of 17 miRNAs between the two categories, with 4 upregulated and 13 down-regulated (including miR-425) miRNAs in male breast cancers [Fassan *et al.* 2009].

MIR-429

miR-429 is a member of the miR-200 family of miRNAs. For more information, see miR-200 family.

MIR-450

E_2 was found to induce miR-450 expression in BCC [Castellano et al. 2009].

MIR-451

Many chemotherapy regimens are successfully used to treat breast cancer; however, often BCC develop drug resistance that usually leads to a relapse and worsening of prognosis. miR-451 was found to regulate the expression of multidrug resistance 1 gene in MCF-7 BCC. Transfection of a doxorubicin-resistant MCF-7 subline with miR-451 resulted in the increased sensitivity of these cells to doxorubicin, suggesting that correction of altered expression of miRNA may have significant implications for therapeutic strategies aiming to overcome cancer cell resistance [Kovalchuk et al. 2008].

MIR-489

A miRNA microarray analysis of MCF-7 BCC lines that are either sensitive (parental) or resistant to tamoxifen showed marked down-regulation of miR-489 miRNA in resistant cells compared with parental MCF-7 BCC [Miller et al. 2008].

MIR-497

In an analysis involving 435 miRNA oligonucleotide probes, miR-497 was found down-regulated in breast cancer, as compared to normal adjacent tumor tissues [Yan et al. 2008].

MIR-499

In a case-control study of 1,009 breast cancer cases and 1,093 cancer-free controls in a population, the miR-499 rs3746444:A>G single nucleotide polymorphism (SNP) was associated with significantly increased risks of breast cancer in a dose-effect manner. This suggests that SNPs in miRNAs may contribute to breast cancer susceptibility [Hu et al. 2009].

In BCC, the expression of a broad set of miRNAs, including miR-499, was decreased following E_2 treatment in an ER-dependent manner [Maillot et al. 2009].

MIR-506

E_2 was found to repress seven miRNAs (miR-302b, miR-506, miR-524, miR-27a, miR-27b, miR-143, miR-9) in MCF-7 BCC [Bhat-Nakshatri et al. 2009].

EZH2 is a multifunctional protein that integrates Wnt and E_2 signaling in BCC and the abnormal function of this protein is linked to several diseases (Shi et al. 2007). Based on TargetScan analysis, EZH2 was identified as a likely target of miR-124a and miR-506, both of which are repressed by E_2 (in [Bhat-Nakshatri et al. 2009]).

MIR-516

miR-516a-3p expression has been associated to ER-positive/LN-negative breast cancer aggressiveness. Bioinformatic analysis coupled miR-516a-3p to cell cycle progression and chromosomal instability [Foekens et al. 2008].

MIR-520

See also miR-373

In BCC, miR-520g was part of a miRNA signature corresponding with progesterone receptor status identified by stepwise artificial neural networks analysis. Its expression was further analyzed in 95 breast tumors. MiR-520g level was elevated in ER and PR-negative tumors [Lowery et al. 2009].

In BCC, miR-520d was part of a miRNA signature corresponding with HER2/neu receptor status identified by stepwise artificial neural networks analysis [Lowery *et al.* 2009]. The expression of a broad set of miRNAs, including miR-520d, was decreased following E_2 treatment of BCC in an ER-α-dependent manner [Maillot *et al.* 2009].

MIR-524

E_2 was found to repress seven miRNAs (miR-302b, miR-506, miR-524, miR-27a, miR-27b, miR-143, miR-9) in MCF-7 BCC [Bhat-Nakshatri *et al.* 2009].

MIR-605

A male/female breast cancer comparison showed a different expression of 17 miRNAs between the two categories, with 4 upregulated (including miR-605) and 13 down-regulated miRNAs in male breast cancers [Fassan *et al.* 2009].

MIR-616

A male/female breast cancer comparison showed a different expression of 17 miRNAs between the two categories, with 4 upregulated (including miR-616) and 13 down-regulated miRNAs in male breast cancers [Fassan *et al.* 2009].

MIR-618

A male/female breast cancer comparison showed a different expression of 17 miRNAs between the two categories, with 4 upregulated (including miR-618) and 13 down-regulated miRNAs in male breast cancers [Fassan *et al.* 2009].

MIR-661

The c/EBPα transcription factor positively regulates miR-661 expression, through direct interaction with the miR-661 putative promoter that contains a c/EBPα-consensus motif. In turn, miR-661 inhibits the expression of metastatic tumor antigen 1 (MTA1), a widely up-regulated gene product in human cancer, by targeting the 3' UTR of MTA1 mRNA. The level of MTA1 protein was progressively up-regulated, whereas that of miR-661 and its activator, c/EBPα, were down-regulated in a breast cancer progression model consisting of MCF-10A cell lines whose phenotypes ranged from noninvasive to highly invasive. c/EBPα expression in BCC resulted in increased miR-661 expression and reduced MTA1 3'UTR-luciferase activity and MTA1 protein level. Finally, the introduction of miR-661 inhibited the motility, invasiveness, anchorage-independent growth, and tumorigenicity of invasive BCC. This suggests a therapeutic use of miR-661 to down-regulate the expression of MTA1 in cancer cells [Reddy et al. 2009].

MIR-663

Aberrant hypermethylation was shown for five miRNAs, including miR-663, in 34-86% of cases in a series of 71 primary human breast cancer specimens [Lehmann *et al.* 2008].

A male/female breast cancer comparison showed a different expression of 17 miRNAs between the two categories, with 4 upregulated (including miR-663) and 13 down-regulated miRNAs in male breast cancers [Fassan *et al.* 2009].

REFERENCES

Adams BD, Furneaux H, White BA. The micro-ribonucleic acid (miRNA) miR-206 targets the human estrogen receptor-alpha (ERalpha) and represses ERalpha messenger RNA and protein expression in breast cancer cell lines. *Mol. Endocrinol.* 2007 May;21(5):1132-47.

Adams BD, Claffey KP, White BA. Argonaute-2 expression is regulated by epidermal growth factor receptor and mitogen-activated protein kinase signaling and correlates with a transformed phenotype in breast cancer cells. *Endocrinology.* 2009 Jan;150(1):14-23.

Aguda BD, Kim Y, Piper-Hunter MG, Friedman A, Marsh CB. MicroRNA regulation of a cancer network: consequences of the feedback loops involving miR-17-92, E2F, and Myc. *Proc. Natl. Acad. Sci. U S A.* 2008 Dec 16;105(50):19678-83.

Amaral FC, Torres N, Saggioro F, Neder L, Machado HR, Silva WA Jr, Moreira AC, Castro M. MicroRNAs differentially expressed in ACTH-secreting pituitary tumors. *J. Clin. Endocrinol. Metab.* 2009 Jan;94(1):320-3.

Bertucci F, Finetti P, Cervera N, Charafe-Jauffret E, Buttarelli M, Jacquemier J, Chaffanet M, Maraninchi D, Viens P, Birnbaum D. How different are luminal A and basal breast cancers? *Int. J. Cancer.* 2009 Mar 15;124(6):1338-48.

Bhat-Nakshatri P, Wang G, Collins NR, Thomson MJ, Geistlinger TR, Carroll JS, Brown M, Hammond S, Srour EF, Liu Y, Nakshatri H. Estradiol-regulated microRNAs control estradiol response in breast cancer cells. *Nucleic Acids Res.* 2009 Aug;37(14):4850-61.

Bhaumik D, Scott GK, Schokrpur S, Patil CK, Campisi J, Benz CC. Expression of microRNA-146 suppresses NF-kappaB activity with reduction of metastatic potential in breast cancer cells. *Oncogene.* 2008 Sep 18;27(42):5643-7.

Blenkiron C, Goldstein LD, Thorne NP, Spiteri I, Chin SF, Dunning MJ, Barbosa-Morais NL, Teschendorff AE, Green AR, Ellis IO, Tavaré S, Caldas C, Miska EA. MicroRNA expression profiling of human breast cancer identifies new markers of tumor subtype. *Genome Biol.* 2007;8(10):R214.

Bonci D, Coppola V, Musumeci M, Addario A, Giuffrida R, Memeo L, D'Urso L, Pagliuca A, Biffoni M, Labbaye C, Bartucci M, Muto G, Peschle C, De Maria R. The miR-15a-miR-16-1 cluster controls prostate cancer by targeting multiple oncogenic activities. *Nat. Med.* 2008 Nov;14(11):1271-7.

Bottoni A, Piccin D, Tagliati F, Luchin A, Zatelli MC, degli Uberti EC. miR-15a and miR-16-1 down-regulation in pituitary adenomas. *J. Cell Physiol.* 2005 Jul;204(1):280-5.

Bourdeau V, Deschênes J, Laperrière D, Aid M, White JH, Mader S. Mechanisms of primary and secondary estrogen target gene regulation in breast cancer cells. *Nucleic Acids Res.* 2008 Jan;36(1):76-93.

Bourguignon LY, Spevak CC, Wong G, Xia W, Gilad E. Hyaluronan-CD44 interaction with PKC-epsilon promotes oncogenic signaling by the stem cell marker, Nanog and the production of microRNA-21 leading to downregulation of the tumor suppressor protein, PDCD4, anti-apoptosis and chemotherapy resistance in breast tumor cells. *J. Biol. Chem.* 2009 Sep 25;284(39):26533-46.

Bracken CP, Gregory PA, Kolesnikoff N, Bert AG, Wang J, Shannon MF, Goodall GJ. A double-negative feedback loop between ZEB1-SIP1 and the microRNA-200 family regulates epithelial-mesenchymal transition. *Cancer Res.* 2008 Oct 1;68(19):7846-54.

Burk U, Schubert J, Wellner U, Schmalhofer O, Vincan E, Spaderna S, Brabletz T. A reciprocal repression between ZEB1 and members of the miR-200 family promotes EMT and invasion in cancer cells. *EMBO Rep.* 2008 Jun;9(6):582-9.

Büssing I, Slack FJ, Grosshans H. let-7 microRNAs in development, stem cells and cancer. *Trends Mol. Med.* 2008 Sep;14(9):400-9.

Calin GA, Cimmino A, Fabbri M, Ferracin M, Wojcik SE, Shimizu M, Taccioli C, Zanesi N, Garzon R, Aqeilan RI, Alder H, Volinia S, Rassenti L, Liu X, Liu CG, Kipps TJ, Negrini M, Croce CM. MiR-15a and miR-16-1 cluster functions in human leukemia. *Proc. Natl. Acad. Sci. U S A.* 2008 Apr 1;105(13):5166-71.

Camps C, Buffa FM, Colella S, Moore J, Sotiriou C, Sheldon H, Harris AL, Gleadle JM, Ragoussis J. hsa-miR-210 Is induced by hypoxia and is an independent prognostic factor in breast cancer. *Clin. Cancer Res.* 2008 Mar 1;14(5):1340-8.

Cao Q, Yu J, Dhanasekaran SM, Kim JH, Mani RS, Tomlins SA, Mehra R, Laxman B, Cao X, Yu J, Kleer CG, Varambally S, Chinnaiyan AM. Repression of E-cadherin by the polycomb group protein EZH2 in cancer. *Oncogene.* 2008 Dec 11;27(58):7274-84.

Castellano L, Giamas G, Jacob J, Coombes RC, Lucchesi W, Thiruchelvam P, Barton G, Jiao LR, Wait R, Waxman J, Hannon GJ, Stebbing J. The estrogen receptor-alpha-induced microRNA signature regulates itself and its transcriptional response. *Proc. Natl. Acad. Sci. U S A.* 2009 Sep 15;106(37):15732-7.

Chan BT, Lee AV. Insulin receptor substrates (IRSs) and breast tumorigenesis. *J. Mammary Gland Biol. Neoplasia.* 2008 Dec;13(4):415-22.

Charafe-Jauffret E, Monville F, Ginestier C, Dontu G, Birnbaum D, Wicha MS. Cancer stem cells in breast: current opinion and future challenges. *Pathobiology.* 2008;75(2):75-84.

Chen YT, Kitabayashi N, Zhou XK, Fahey TJ 3rd, Scognamiglio T. MicroRNA analysis as a potential diagnostic tool for papillary thyroid carcinoma. *Mod. Pathol.* 2008 Sep;21(9):1139-46.

Chuthapisith S, Bean BE, Cowley G, Eremin JM, Samphao S, Layfield R, Kerr ID, Wiseman J, El-Sheemy M, Sreenivasan T, Eremin O. Annexins in human breast cancer: Possible predictors of pathological response to neoadjuvant chemotherapy. *Eur. J. Cancer.* 2009 May;45(7):1274-81.

Cimmino A, Calin GA, Fabbri M, Iorio MV, Ferracin M, Shimizu M, Wojcik SE, Aqeilan RI, Zupo S, Dono M, Rassenti L, Alder H, Volinia S, Liu CG, Kipps TJ, Negrini M, Croce CM. miR-15 and miR-16 induce apoptosis by targeting BCL2. *Proc. Natl. Acad. Sci. U S A.* 2005 Sep 27;102 (39):13944-9.

Cloonan N, Brown MK, Steptoe AL, Wani S, Chan WL, Forrest AR, Kolle G, Gabrielli B, Grimmond SM. The miR-17-5p microRNA is a key regulator of the G1/S phase cell cycle transition. *Genome Biol.* 2008;9(8):R127.

Cochrane DR, Spoelstra NS, Howe EN, Nordeen SK, Richer JK. MicroRNA-200c mitigates invasiveness and restores sensitivity to microtubule-targeting chemo therapeutic agents. Mol Cancer Ther. 2009 May 12. [Epub ahead of print]

Crawford M, Brawner E, Batte K, Yu L, Hunter MG, Otterson GA, Nuovo G, Marsh CB, Nana-Sinkam SP. MicroRNA-126 inhibits invasion in non-small cell lung carcinoma cell lines. *Biochem. Biophys. Res. Commun.* 2008 Sep 5;373(4):607-12.

Crosby ME, Kulshreshtha R, Ivan M, Glazer PM. MicroRNA regulation of DNA repair gene expression in hypoxic stress. *Cancer Res.* 2009 Feb 1;69(3):1221-9.

Cuesta R, Martínez-Sánchez A, Gebauer F. miR-181a regulates cap-dependent translation of p27kip1 mRNA in myeloid cells. *Mol. Cell Biol.* 2009 May;29(10):2841-51.

Davis BN, Hilyard AC, Lagna G, Hata A. SMAD proteins control DROSHA-mediated microRNA maturation. *Nature.* 2008 Jul 3;454(7200):56-61.

Dews M, Homayouni A, Yu D, Murphy D, Sevignani C, Wentzel E, Furth EE, Lee WM, Enders GH, Mendell JT, Thomas-Tikhonenko A. Augmentation of tumor angiogenesis by a Myc-activated microRNA cluster. *Nat Genet.* 2006 Sep;38(9):1060-5.

Faraoni I, Antonetti FR, Cardone J, Bonmassar E. miR-155 gene: A typical multifunctional microRNA. *Biochim Biophys Acta.* 2009 Jun;1792(6):497-505.

Fassan M, Baffa R, Palazzo JP, Lloyd J, Crosariol M, Liu CG, Volinia S, Alder H, Rugge M, Croce CM, Rosenberg A. MicroRNA expression profiling of male breast cancer. *Breast Cancer Res.* 2009;11(4):R58.

Felicetti F, Errico MC, Bottero L, Segnalini P, Stoppacciaro A, Biffoni M, Felli N, Mattia G, Petrini M, Colombo MP, Peschle C, Carè A. The promyelocytic leukemia zinc finger-microRNA-221/-222 pathway controls melanoma progression through multiple oncogenic mechanisms. *Cancer Res.* 2008 Apr 15;68(8):2745-54.

Ferretti E, De Smaele E, Po A, Di Marcotullio L, Tosi E, Espinola MS, Di Rocco C, Riccardi R, Giangaspero F, Farcomeni A, Nofroni I, Laneve P, Gioia U, Caffarelli E, Bozzoni I, Screpanti I, Gulino A. MicroRNA profiling in human medulloblastoma. *Int. J. Cancer.* 2009 Feb 1;124(3):568-77.

Foekens JA, Sieuwerts AM, Smid M, Look MP, de Weerd V, Boersma AW, Klijn JG, Wiemer EA, Martens JW. Four miRNAs associated with aggressiveness of lymph node-negative, estrogen receptor-positive human breast cancer. *Proc. Natl. Acad. Sci. U S A.* 2008 Sep 2;105(35):13021-6.

Frankel LB, Christoffersen NR, Jacobsen A, Lindow M, Krogh A, Lund AH. Programmed cell death 4 (PDCD4) is an important functional target of the microRNA miR-21 in breast cancer cells. *J. Biol. Chem.* 2008 Jan 11;283(2):1026-33.

Fulci V, Chiaretti S, Goldoni M, Azzalin G, Carucci N, Tavolaro S, Castellano L, Magrelli A, Citarella F, Messina M, Maggio R, Peragine N, Santangelo S, Mauro FR, Landgraf P, Tuschl T, Weir DB, Chien M, Russo JJ, Ju J, Sheridan R, Sander C, Zavolan M, Guarini A, Foà R, Macino G. Quantitative technologies establish a novel microRNA profile of chronic lymphocytic leukemia. *Blood.* 2007 Jun 1;109(11):4944-51.

Galardi S, Mercatelli N, Giorda E, Massalini S, Frajese GV, Ciafrè SA, Farace MG. miR-221 and miR-222 expression affects the proliferation potential of human prostate carcinoma cell lines by targeting p27Kip1. *J. Biol. Chem.* 2007 Aug 10;282(32):23716-24.

Gee HE, Camps C, Buffa FM, Colella S, Sheldon H, Gleadle JM, Ragoussis J, Harris AL. MicroRNA-10b and breast cancer metastasis. *Nature.* 2008 Oct 23;455(7216):E8-9.

Giannakakis A, Sandaltzopoulos R, Greshock J, Liang S, Huang J, Hasegawa K, Li C, O'Brien-Jenkins A, Katsaros D, Weber BL, Simon C, Coukos G, Zhang L. miR-210 links hypoxia with cell cycle regulation and is deleted in human epithelial ovarian cancer. *Cancer Biol. Ther.* 2008 Feb;7(2):255-64.

Gregory PA, Bert AG, Paterson EL, Barry SC, Tsykin A, Farshid G, Vadas MA, Khew-Goodall Y, Goodall GJ. The miR-200 family and miR-205 regulate epithelial to mesenchymal transition by targeting ZEB1 and SIP1. *Nat. Cell Biol.* 2008 May;10(5):593-601.

Guo C, Sah JF, Beard L, Willson JK, Markowitz SD, Guda K. The noncoding RNA, miR-126, suppresses the growth of neoplastic cells by targeting phosphatidylinositol 3-kinase signaling and is frequently lost in colon cancers. Genes Chromosomes *Cancer.* 2008 Nov;47(11):939-46.

Guo X, Wu Y, Hartley RS. MicroRNA-125a represses cell growth by targeting HuR in breast cancer. *RNA Biol.* 2009 Nov-Dec;6(5):575-83.

Guttilla IK, White BA. Coordinate Regulation of FOXO1 by miR-27a, miR-96, and miR-182 in Breast Cancer Cells. *J. Biol. Chem.* 2009 Aug 28;284(35):23204-16.

Habbe N, Koorstra JB, Mendell JT, Offerhaus GJ, Ryu JK, Feldmann G, Mullendore ME, Goggins MG, Hong SM, Maitra A. MicroRNA miR-155 is a biomarker of early pancreatic neoplasia. *Cancer Biol. Ther.* 2009 Feb;8(4):340-6.

He H, Jazdzewski K, Li W, Liyanarachchi S, Nagy R, Volinia S, Calin GA, Liu CG, Franssila K, Suster S, Kloos RT, Croce CM, de la Chapelle A. The role of microRNA genes in papillary thyroid carcinoma. *Proc Natl Acad. Sci. U S A.* 2005 Dec 27;102(52):19075-80.

He X, He L, Hannon GJ. The guardian's little helper: microRNAs in the p53 tumor suppressor network. *Cancer Res.* 2007 Dec 1;67(23):11099-101.

Helms MW, Packeisen J, August C, Schittek B, Boecker W, Brandt BH, Buerger H. First evidence supporting a potential role for the BMP/SMAD pathway in the progression of oestrogen receptor-positive breast cancer. *J. Pathol.* 2005 Jul;206(3):366-76.

Hermeking H. p53 enters the microRNA world. *Cancer Cell.* 2007 Nov;12(5):414-8.

Hoffman AE, Zheng T, Yi C, Leaderer D, Weidhaas J, Slack F, Zhang Y, Paranjape T, Zhu Y. microRNA miR-196a-2 and breast cancer: a genetic and epigenetic association study and functional analysis. *Cancer Res.* 2009 Jul 15;69(14):5970-7.

Hoover LL, Kubalak SW. Holding their own: the noncanonical roles of Smad proteins. *Sci Signal.* 2008 Nov 18;1(46):pe48.

Hossain A, Kuo MT, Saunders GF. Mir-17-5p regulates breast cancer cell proliferation by inhibiting translation of AIB1 mRNA. *Mol. Cell Biol.* 2006 Nov;26(21):8191-201.

Hsu PY, Deatherage DE, Rodriguez BA, Liyanarachchi S, Weng YI, Zuo T, Liu J, Cheng AS, Huang TH. Xenoestrogen-induced epigenetic repression of microRNA-9-3 in breast epithelial cells. *Cancer Res.* 2009 Jul 15;69(14):5936-45.

Hu Z, Liang J, Wang Z, Tian T, Zhou X, Chen J, Miao R, Wang Y, Wang X, Shen H. Common genetic variants in pre-microRNAs were associated with increased risk of breast cancer in Chinese women. *Hum. Mutat.* 2009 Jan;30(1):79-84.

Huang Q, Gumireddy K, Schrier M, le Sage C, Nagel R, Nair S, Egan DA, Li A, Huang G, Klein-Szanto AJ, Gimotty PA, Katsaros D, Coukos G, Zhang L, Puré E, Agami R. The microRNAs miR-373 and miR-520c promote tumour invasion and metastasis. *Nat Cell Biol.* 2008 Feb;10(2):202-10.

Huang TH, Wu F, Loeb GB, Hsu R, Heidersbach A, Brincat A, Horiuchi D, Lebbink RJ, Mo YY, Goga A, McManus MT. Up-regulation of miR-21 by HER2/neu Signaling Promotes Cell Invasion. *J. Biol Chem.* 2009 Jul 3;284(27):18515-24.

Hurst DR, Edmonds MD, Scott GK, Benz CC, Vaidya KS, Welch DR. Breast cancer metastasis suppressor 1 up-regulates miR-146, which suppresses breast cancer metastasis. *Cancer Res.* 2009 Feb 15;69(4):1279-83.

Hurteau GJ, Carlson JA, Spivack SD, Brock GJ. Overexpression of the microRNA hsa-miR-200c leads to reduced expression of transcription factor 8 (ZEB1) and increased expression of E-cadherin. *Cancer Res.* 2007 Sep 1;67(17):7972-6.

Hwang C, Giri VN, Wilkinson JC, Wright CW, Wilkinson AS, Cooney KA, Duckett CS. EZH2 regulates the transcription of estrogen-responsive genes through association with REA, an estrogen receptor corepressor. *Breast Cancer Res. Treat.* 2008 Jan;107(2):235-42.

Iliopoulos D, Hirsch HA, Struhl K. An Epigenetic Switch Involving NF-kappaB, Lin28, Let-7 MicroRNA, and IL6 Links Inflammation to Cell Transformation. *Cell.* 2009 Nov 13;139(4):693-706.

Iliopoulos D, Polytarchou C, Hatziapostolou M, Kottakis F, Maroulakou IG, Struhl K, Tsichlis PN. MicroRNAs differentially regulated by Akt isoforms control EMT and stem cell renewal in cancer cells. *Sci Signal.* 2009b Oct 13;2(92):ra62.

Iorio MV, Ferracin M, Liu CG, Veronese A, Spizzo R, Sabbioni S, Magri E, Pedriali M, Fabbri M, Campiglio M, Ménard S, Palazzo JP, Rosenberg A, Musiani P, Volinia S, Nenci I, Calin GA, Querzoli P, Negrini M, Croce CM. MicroRNA gene expression deregulation in human breast cancer. *Cancer Res.* 2005 Aug 15;65(16):7065-70.

Iorio MV, Visone R, Di Leva G, Donati V, Petrocca F, Casalini P, Taccioli C, Volinia S, Liu CG, Alder H, Calin GA, Ménard S, Croce CM. MicroRNA signatures in human ovarian cancer. *Cancer Res.* 2007 Sep 15;67(18): 8699-707.

Iorio MV, Casalini P, Piovan C, Di Leva G, Merlo A, Triulzi T, Ménard S, Croce CM, Tagliabue E. microRNA-205 Regulates HER3 in Human Breast Cancer. *Cancer Res.* 2009 Mar 15;69(6):2195-200.

Ivan M, Harris AL, Martelli F, Kulshreshtha R. Hypoxia response and micro RNAs: no longer two separate worlds. *J. Cell Mol. Med.* 2008 Sep-Oct;12 (5A):1426-31.

Jung M, Mollenkopf HJ, Grimm C, Wagner I, Albrecht M, Waller T, Pilarsky C, Johannsen M, Stephan C, Lehrach H, Nietfeld W, Rudel T, Jung K, Kristiansen G. MicroRNA profiling of clear cell renal cell cancer identifies a robust signature to define renal malignancy. *J. Cell Mol. Med.* 2009 Feb 17. [Epub ahead of print]

Kato M, Paranjape T, Ullrich R, Nallur S, Gillespie E, Keane K, Esquela-Kerscher A, Weidhaas JB, Slack FJ. The mir-34 microRNA is required for the DNA damage response in vivo in C. elegans and in vitro in human breast cancer cells. *Oncogene.* 2009 Jun 25;28(25):2419-24.

Kefas B, Godlewski J, Comeau L, Li Y, Abounader R, Hawkinson M, Lee J, Fine H, Chiocca EA, Lawler S, Purow B. microRNA-7 inhibits the epidermal growth factor receptor and the Akt pathway and is down-regulated in glioblastoma. *Cancer Res.* 2008 May 15;68(10):3566-72.

Kim YK, Yu J, Han TS, Park SY, Namkoong B, Kim DH, Hur K, Yoo MW, Lee HJ, Yang HK, Kim VN. Functional links between clustered microRNAs: suppression of cell-cycle inhibitors by microRNA clusters in gastric cancer. *Nucleic Acids Res.* 2009 Apr;37(5):1672-81.

Kleer CG, Griffith KA, Sabel MS, Gallagher G, van Golen KL, Wu ZF, Merajver SD. RhoC-GTPase is a novel tissue biomarker associated with biologically aggressive carcinomas of the breast. *Breast Cancer Res Treat.* 2005 Sep;93(2):101-10.

Komagata S, Nakajima M, Takagi S, Mohri T, Taniya T, Yokoi T. Human CYP24 catalyzing the inactivation of calcitriol is post-transcriptionally regulated by miR-125b. *Mol. Pharmacol.* 2009 Oct;76(4):702-9.

Kondo N, Toyama T, Sugiura H, Fujii Y, Yamashita H. miR-206 Expression is down-regulated in estrogen receptor alpha-positive human breast cancer. *Cancer Res.* 2008 Jul 1;68(13):5004-8.

Kong W, Yang H, He L, Zhao JJ, Coppola D, Dalton WS, Cheng JQ. MicroRNA-155 is regulated by the TGF-β/Smad pathway and contributes to epithelial cell plasticity by targeting RhoA. *Mol Cell Biol.* 2008 Nov;28(22):6773-84.

Korpal M, Lee ES, Hu G, Kang Y. The miR-200 family inhibits epithelial-mesenchymal transition and cancer cell migration by direct targeting of E-cadherin transcriptional repressors ZEB1 and ZEB2. *J. Biol. Chem.* 2008 May 30;283(22):14910-4.

Korpal M, Kang Y. The emerging role of miR-200 family of microRNAs in epithelial-mesenchymal transition and cancer metastasis. *RNA Biol.* 2008 Jul-Sep;5(3):115-9.

Kovalchuk O, Filkowski J, Meservy J, Ilnytskyy Y, Tryndyak VP, Chekhun VF, Pogribny IP. Involvement of microRNA-451 in resistance of the MCF-7 breast cancer cells to chemotherapeutic drug doxorubicin. *Mol. Cancer Ther.* 2008 Jul;7(7):2152-9.

Krek A, Grün D, Poy MN, Wolf R, Rosenberg L, Epstein EJ, MacMenamin P, da Piedade I, Gunsalus KC, Stoffel M, Rajewsky N. Combinatorial microRNA target predictions. *Nat Genet.* 2005 May;37(5):495-500.

Kuehbacher A, Urbich C, Dimmeler S. Targeting microRNA expression to regulate angiogenesis. *Trends Pharmacol Sci.* 2008 Jan;29(1):12-5.

Kulshreshtha R, Ferracin M, Wojcik SE, Garzon R, Alder H, Agosto-Perez FJ, Davuluri R, Liu CG, Croce CM, Negrini M, Calin GA, Ivan M. A microRNA signature of hypoxia. *Mol. Cell Biol.* 2007 Mar;27(5):1859-67.

Lacroix M, Leclercq G. Relevance of breast cancer cell lines as models for breast tumours: an update. *Breast Cancer Res Treat.* 2004a Feb;83(3): 249-89.

Lacroix M, Leclercq G. About GATA3, HNF3A, and XBP1, three genes co-expressed with the oestrogen receptor-alpha gene (ESR1) in breast cancer. *Mol. Cell Endocrinol.* 2004b Apr 30;219(1-2):1-7.

Lacroix M, Toillon RA, Leclercq G. Stable 'portrait' of breast tumors during progression: data from biology, pathology and genetics. *Endocr. Relat. Cancer.* 2004 Sep;11(3):497-522.

Lacroix M, Toillon RA, Leclercq G. p53 and breast cancer, an update. *Endocr. Relat. Cancer.* 2006 Jun;13(2):293-325.

Lacroix M. MDA-MB-435 cells are from melanoma, not from breast cancer. *Cancer Chemother. Pharmacol.* 2009 Feb;63(3):567.

Lagos-Quintana M, Rauhut R, Lendeckel W, Tuschl T. Identification of novel genes coding for small expressed RNAs. *Science.* 2001 Oct 26;294(5543):853-8.

Laios A, O'Toole S, Flavin R, Martin C, Kelly L, Ring M, Finn SP, Barrett C, Loda M, Gleeson N, D'Arcy T, McGuinness E, Sheils O, Sheppard B, O'

Leary J. Potential role of miR-9 and miR-223 in recurrent ovarian cancer. *Mol. Cancer.* 2008 Apr 28;7:35.

Landgraf P, Rusu M, Sheridan R, Sewer A, Iovino N, Aravin A, Pfeffer S, Rice A, Kamphorst AO, Landthaler M, Lin C, Socci ND, Hermida L, Fulci V, Chiaretti S, Foà R, Schliwka J, Fuchs U, Novosel A, Müller RU, Schermer B, Bissels U, Inman J, Phan Q, Chien M, Weir DB, Choksi R, De Vita G, Frezzetti D, Trompeter HI, Hornung V, Teng G, Hartmann G, Palkovits M, Di Lauro R, Wernet P, Macino G, Rogler CE, Nagle JW, Ju J, Papavasiliou FN, Benzing T, Lichter P, Tam W, Brownstein MJ, Bosio A, Borkhardt A, Russo JJ, Sander C, Zavolan M, Tuschl T. A mammalian microRNA expression atlas based on small RNA library sequencing. *Cell.* 2007 Jun 29;129(7):1401-14.

Lee EJ, Gusev Y, Jiang J, Nuovo GJ, Lerner MR, Frankel WL, Morgan DL, Postier RG, Brackett DJ, Schmittgen TD. Expression profiling identifies microRNA signature in pancreatic cancer. *Int J. Cancer.* 2007 Mar 1;120(5):1046-54.

Lehmann U, Hasemeier B, Christgen M, Müller M, Römermann D, Länger F, Kreipe H. Epigenetic inactivation of microRNA gene hsa-mir-9-1 in human breast cancer. *J. Pathol.* 2008 Jan;214(1):17-24.

Leivonen SK, Mäkelä R, Ostling P, Kohonen P, Haapa-Paananen S, Kleivi K, Enerly E, Aakula A, Hellström K, Sahlberg N, Kristensen VN, Børresen-Dale AL, Saviranta P, Perälä M, Kallioniemi O. Protein lysate microarray analysis to identify microRNAs regulating estrogen receptor signaling in breast cancer cell lines. *Oncogene.* 2009 Nov 5;28(44):3926-36.

le Sage C, Nagel R, Egan DA, Schrier M, Mesman E, Mangiola A, Anile C, Maira G, Mercatelli N, Ciafrè SA, Farace MG, Agami R. Regulation of the p27(Kip1) tumor suppressor by miR-221 and miR-222 promotes cancer cell proliferation. *EMBO J.* 2007 Aug 8;26(15):3699-708.

Lewis BP, Burge CB, Bartel DP. Conserved seed pairing, often flanked by adenosines, indicates that thousands of human genes are microRNA targets. *Cell.* 2005 Jan 14;120(1):15-20.

Li W, Duan R, Kooy F, Sherman SL, Zhou W, Jin P. Germline mutation of microRNA-125a is associated with breast cancer. *J. Med. Genet.* 2009a May;46(5):358-60.

Li XF, Yan PJ, Shao ZM. Downregulation of miR-193b contributes to enhance urokinase-type plasminogen activator (uPA) expression and tumor progression and invasion in human breast cancer. *Oncogene.* 2009b Nov 5;28(44):3937-48.

Liang Z, Wu H, Xia J, Li Y, Zhang Y, Huang K, Wagar N, Yoon Y, Cho HT, Scala S, Shim H. Involvement of miR-326 in chemotherapy resistance of breast cancer through modulating expression of multidrug resistance-associated protein 1. *Biochem Pharmacol.* 2010 Mar 15;79(6):817-24.

Liu B, Peng XC, Zheng XL, Wang J, Qin YW. MiR-126 restoration downregulate VEGF and inhibit the growth of lung cancer cell lines in vitro and in vivo. *Lung Cancer.* 2009 Nov;66(2):169-75.

Lodygin D, Tarasov V, Epanchintsev A, Berking C, Knyazeva T, Körner H, Knyazev P, Diebold J, Hermeking H. Inactivation of miR-34a by aberrant CpG methylation in multiple types of cancer. *Cell Cycle.* 2008 Aug 15;7(16):2591-600.

Lowery AJ, Miller N, Devaney A, McNeill RE, Davoren PA, Lemetre C, Benes V, Schmidt S, Blake J, Ball G, Kerin MJ. MicroRNA signatures predict oestrogen receptor, progesterone receptor and HER2/neu receptor status in breast cancer. *Breast Cancer Res.* 2009;11(3):R27.

Lu Z, Liu M, Stribinskis V, Klinge CM, Ramos KS, Colburn NH, Li Y. MicroRNA-21 promotes cell transformation by targeting the programmed cell death 4 gene. *Oncogene.* 2008 Jul 17;27(31):4373-9.

Lujambio A, Esteller M. CpG island hypermethylation of tumor suppressor microRNAs in human cancer. *Cell Cycle.* 2007 Jun 15;6(12):1455-9.

Lujambio A, Calin GA, Villanueva A, Ropero S, Sánchez-Céspedes M, Blanco D, Montuenga LM, Rossi S, Nicoloso MS, Faller WJ, Gallagher WM, Eccles SA, Croce CM, Esteller M. A microRNA DNA methylation signature for human cancer metastasis. *Proc. Natl. Acad. Sci U S A.* 2008 Sep 9;105(36):13556-61.

Luthra R, Singh RR, Luthra MG, Li YX, Hannah C, Romans AM, Barkoh BA, Chen SS, Ensor J, Maru DM, Broaddus RR, Rashid A, Albarracin CT. MicroRNA-196a targets annexin A1: a microRNA-mediated mechanism of annexin A1 downregulation in cancers. *Oncogene.* 2008 Nov 6;27(52):6667-78.

Luzi E, Marini F, Sala SC, Tognarini I, Galli G, Brandi ML. Osteogenic differentiation of human adipose tissue-derived stem cells is modulated by the miR-26a targeting of the SMAD1 transcription factor. *J. Bone Miner Res.* 2008 Feb;23(2):287-95.

Ma L, Teruya-Feldstein J, Weinberg RA. Tumour invasion and metastasis initiated by microRNA-10b in breast cancer. *Nature.* 2007 Oct 11;449(7163):682-8.

Maillot G, Lacroix-Triki M, Pierredon S, Gratadou L, Schmidt S, Bénès V, Roché H, Dalenc F, Auboeuf D, Millevoi S, Vagner S. Widespread

Estrogen-Dependent Repression of microRNAs Involved in Breast Tumor Cell Growth. *Cancer Res.* 2009 Nov 1;69(21):8332-40.

Mattie MD, Benz CC, Bowers J, Sensinger K, Wong L, Scott GK, Fedele V, Ginzinger D, Getts R, Haqq C. Optimized high-throughput microRNA expression profiling provides novel biomarker assessment of clinical prostate and breast cancer biopsies. *Mol Cancer.* 2006 Jun 19;5:24.

Medina R, Zaidi SK, Liu CG, Stein JL, van Wijnen AJ, Croce CM, Stein GS. MicroRNAs 221 and 222 bypass quiescence and compromise cell survival. *Cancer Res.* 2008 Apr 15;68(8):2773-80.

Mendell JT. miRiad roles for the miR-17-92 cluster in development and disease. *Cell.* 2008 Apr 18;133(2):217-22.

Mercatelli N, Coppola V, Bonci D, Miele F, Costantini A, Guadagnoli M, Bonanno E, Muto G, Frajese GV, De Maria R, Spagnoli LG, Farace MG, Ciafrè SA. The inhibition of the highly expressed miR-221 and miR-222 impairs the growth of prostate carcinoma xenografts in mice. *PLoS ONE.* 2008;3(12):e4029.

Mertens-Talcott SU, Chintharlapalli S, Li X, Safe S. The oncogenic microRNA-27a targets genes that regulate specificity protein transcription factors and the G2-M checkpoint in MDA-MB-231 breast cancer cells. *Cancer Res.* 2007 Nov 15;67(22):11001-11.

Miller TE, Ghoshal K, Ramaswamy B, Roy S, Datta J, Shapiro CL, Jacob S, Majumder S. MicroRNA-221/222 confers tamoxifen resistance in breast cancer by targeting p27Kip1. *J. Biol. Chem.* 2008 Oct 31;283(44):29897-903.

Mourelatos Z, Dostie J, Paushkin S, Sharma A, Charroux B, Abel L, Rappsilber J, Mann M, Dreyfuss G. miRNPs: a novel class of ribonucleoproteins containing numerous microRNAs. *Genes Dev.* 2002 Mar 15;16(6):720-8.

Nikiforova MN, Tseng GC, Steward D, Diorio D, Nikiforov YE. MicroRNA expression profiling of thyroid tumors: biological significance and diagnostic utility. *J. Clin. Endocrinol. Metab.* 2008 May;93(5):1600-8.

Omura N, Li CP, Li A, Hong SM, Walter K, Jimeno A, Hidalgo M, Goggins M. Genome-wide profiling of methylated promoters in pancreatic adenocarcinoma. *Cancer Biol. Ther.* 2008 Jul;7(7):1146-56.

Ou K, Yu K, Kesuma D, Hooi M, Huang N, Chen W, Lee SY, Goh XP, Tan LK, Liu J, Soon SY, Bin Abdul Rashid S, Putti TC, Jikuya H, Ichikawa T, Nishimura O, Salto-Tellez M, Tan P. Novel breast cancer biomarkers identified by integrative proteomic and gene expression mapping. *J. Proteome Res.* 2008 Apr;7(4):1518-28.

Ozen M, Creighton CJ, Ozdemir M, Ittmann M. Widespread deregulation of microRNA expression in human prostate cancer. *Oncogene.* 2008 Mar 13;27(12):1788-93.

Pallante P, Visone R, Ferracin M, Ferraro A, Berlingieri MT, Troncone G, Chiappetta G, Liu CG, Santoro M, Negrini M, Croce CM, Fusco A. MicroRNA deregulation in human thyroid papillary carcinomas. *Endocr Relat. Cancer.* 2006 Jun;13(2):497-508.

Pan YZ, Morris ME, Yu AM. MicroRNA-328 negatively regulates the expression of breast cancer resistance protein (BCRP/ABCG2) in human cancer cells. *Mol. Pharmacol.* 2009 Jun;75(6):1374-9.

Pandey DP, Picard D. miR-22 inhibits estrogen signaling by directly targeting the estrogen receptor alpha mRNA. *Mol. Cell Biol.* 2009 Jul;29(13):3783-90.

Peter ME. Let-7 and miR-200 microRNAs: Guardians against pluripotency and cancer progression. *Cell Cycle.* 2009 Mar 15;8(6):843-52.

Poliseno L, Tuccoli A, Mariani L, Evangelista M, Citti L, Woods K, Mercatanti A, Hammond S, Rainaldi G. MicroRNAs modulate the angiogenic properties of HUVECs. *Blood.* 2006 Nov 1;108(9):3068-71.

Qi L, Bart J, Tan LP, Platteel I, Sluis T, Huitema S, Harms G, Fu L, Hollema H, Berg A. Expression of miR-21 and its targets (PTEN, PDCD4, TM1) in flat epithelial atypia of the breast in relation to ductal carcinoma in situ and invasive carcinoma. *BMC Cancer.* 2009 May 28;9:163.

Qian B, Katsaros D, Lu L, Preti M, Durando A, Arisio R, Mu L, Yu H. High miR-21 expression in breast cancer associated with poor disease-free survival in early stage disease and high TGF-beta1. *Breast Cancer Res Treat.* 2009 Sep;117(1):131-40.

Reddy SD, Ohshiro K, Rayala SK, Kumar R. MicroRNA-7, a homeobox D10 target, inhibits p21-activated kinase 1 and regulates its functions. *Cancer Res.* 2008 Oct 15;68(20):8195-200.

Reddy SD, Pakala SB, Ohshiro K, Rayala SK, Kumar R. MicroRNA-661, a c/EBPalpha target, inhibits metastatic tumor antigen 1 and regulates its functions. *Cancer Res.* 2009 Jul 15;69(14):5639-42.

Roush S, Slack FJ. The let-7 family of microRNAs. *Trends Cell Biol.* 2008 Oct;18(10):505-16.

Sachdeva M, Zhu S, Wu F, Wu H, Walia V, Kumar S, Elble R, Watabe K, Mo YY. p53 represses c-Myc through induction of the tumor suppressor miR-145. *Proc. Natl. Acad. Sci. U S A.* 2009 Mar 3;106(9):3207-12.

Sander S, Bullinger L, Wirth T. Repressing the repressor: A new mode of MYC action in lymphomagenesis *Cell Cycle.* 2009 Feb 15;8(4):556-9.

Schepeler T, Reinert JT, Ostenfeld MS, Christensen LL, Silahtaroglu AN, Dyrskjøt L, Wiuf C, Sørensen FJ, Kruhøffer M, Laurberg S, Kauppinen S, Ørntoft TF, Andersen CL. Diagnostic and prognostic microRNAs in stage II colon cancer. *Cancer Res.* 2008 Aug 1;68(15):6416-24.

Scott GK, Goga A, Bhaumik D, Berger CE, Sullivan CS, Benz CC. Coordinate suppression of ERBB2 and ERBB3 by enforced expression of micro-RNA miR-125a or miR-125b. *J.Biol. Chem.* 2007 Jan 12;282(2):1479-86.

Sempere LF, Christensen M, Silahtaroglu A, Bak M, Heath CV, Schwartz G, Wells W, Kauppinen S, Cole CN. Altered MicroRNA expression confined to specific epithelial cell subpopulations in breast cancer. *Cancer Res.* 2007 Dec 15;67(24):11612-20.

Shen J, Ambrosone CB, DiCioccio RA, Odunsi K, Lele SB, Zhao H. A functional polymorphism in the miR-146a gene and age of familial breast/ovarian cancer diagnosis. *Carcinogenesis.* 2008 Oct;29(10):1963-6.

Shi B, Liang J, Yang X, Wang Y, Zhao Y, Wu H, Sun L, Zhang Y, Chen Y, Li R, Zhang Y, Hong M, Shang Y. Integration of estrogen and Wnt signaling circuits by the polycomb group protein EZH2 in breast cancer cells. *Mol. Cell. Biol.* 2007 Jul;27(14):5105-19.

Shimono Y, Zabala M, Cho RW, Lobo N, Dalerba P, Qian D, Diehn M, Liu H, Panula SP, Chiao E, Dirbas FM, Somlo G, Pera RA, Lao K, Clarke MF. Downregulation of miRNA-200c links breast cancer stem cells with normal stem cells. *Cell.* 2009 Aug 7;138(3):592-603.

Si ML, Zhu S, Wu H, Lu Z, Wu F, Mo YY. miR-21-mediated tumor growth. *Oncogene.* 2007 Apr 26;26(19):2799-803.

Slaby O, Svoboda M, Fabian P, Smerdova T, Knoflickova D, Bednarikova M, Nenutil R, Vyzula R. Altered expression of miR-21, miR-31, miR-143 and miR-145 is related to clinicopathologic features of colorectal cancer. *Oncology.* 2007;72(5-6):397-402.

Sorlie T, Tibshirani R, Parker J, Hastie T, Marron JS, Nobel A, Deng S, Johnsen H, Pesich R, Geisler S, Demeter J, Perou CM, Lønning PE, Brown PO, Børresen-Dale AL, Botstein D. Repeated observation of breast tumor subtypes in independent gene expression data sets. *Proc. Natl. Acad. Sci. U S A.* 2003 Jul 8;100(14):8418-23.

Spizzo R, Nicoloso MS, Lupini L, Lu Y, Fogarty J, Rossi S, Zagatti B, Fabbri M, Veronese A, Liu X, Davuluri R, Croce CM, Mills G, Negrini M, Calin GA. miR-145 participates with TP53 in a death-promoting regulatory loop and targets estrogen receptor-alpha in human breast cancer cells. *Cell Death Differ.* 2010 Feb;17(2):246-54.

Sylvestre Y, De Guire V, Querido E, Mukhopadhyay UK, Bourdeau V, Major F, Ferbeyre G, Chartrand P. An E2F/miR-20a autoregulatory feedback loop. *J. Biol. Chem.* 2007 Jan 26;282(4):2135-43.

Tan Y, Zhang B, Wu T, Skogerbo G, Zhu X, Guo X, He S, Chen R. Transcriptional inhibition of Hoxd4 expression by noncoding RNAs in human breast cancer cells. *BMC Mol. Biol.* 2009 Feb 22;10(1):12.

Tanzer A, Stadler PF. Molecular evolution of a microRNA cluster. *J. Mol. Biol.* 2004 May 28;339(2):327-35.

Tavazoie SF, Alarcón C, Oskarsson T, Padua D, Wang Q, Bos PD, Gerald WL, Massagué J. Endogenous human microRNAs that suppress breast cancer metastasis. *Nature.* 2008 Jan 10;451(7175):147-52.

Tonini T, D'Andrilli G, Fucito A, Gaspa L, Bagella L. Importance of Ezh2 polycomb protein in tumorigenesis process interfering with the pathway of growth suppressive key elements. *J. Cell Physiol.* 2008 Feb;214(2):295-300.

Tsang WP, Kwok TT. Let-7a microRNA suppresses therapeutics-induced cancer cell death by targeting caspase-3. *Apoptosis.* 2008 Oct;13 (10): 1215-22.

Turner M, Vigorito E. Regulation of B- and T-cell differentiation by a single microRNA. *Biochem. Soc .Trans.* 2008 Jun;36(Pt 3):531-3.

Valastyan S, Reinhardt F, Benaich N, Calogrias D, Szász AM, Wang ZC, Brock JE, Richardson AL, Weinberg RA. A pleiotropically acting microRNA, miR-31, inhibits breast cancer metastasis. *Cell.* 2009a Jun 12;137(6):1032-46.

Valastyan S, Benaich N, Chang A, Reinhardt F, Weinberg RA. Concomitant suppression of three target genes can explain the impact of a microRNA on metastasis. *Genes Dev.* 2009 Nov 15;23(22):2592-7.

Varnholt H, Drebber U, Schulze F, Wedemeyer I, Schirmacher P, Dienes HP, Odenthal M. MicroRNA gene expression profile of hepatitis C virus-associated hepatocellular carcinoma. *Hepatology.* 2008 Apr;47(4):1223-32.

Ventura A, Young AG, Winslow MM, Lintault L, Meissner A, Erkeland SJ, Newman J, Bronson RT, Crowley D, Stone JR, Jaenisch R, Sharp PA, Jacks T. Targeted deletion reveals essential and overlapping functions of the miR-17 through 92 family of miRNA clusters. *Cell.* 2008 Mar 7;132(5):875-86.

Visone R, Russo L, Pallante P, De Martino I, Ferraro A, Leone V, Borbone E, Petrocca F, Alder H, Croce CM, Fusco A. MicroRNAs (miR)-221 and miR-222, both overexpressed in human thyroid papillary carcinomas, regulate p27Kip1 protein levels and cell cycle. *Endocr. Relat. Cancer.* 2007 Sep;14(3):791-8.

Volinia S, Calin GA, Liu CG, Ambs S, Cimmino A, Petrocca F, Visone R, Iorio M, Roldo C, Ferracin M, Prueitt RL, Yanaihara N, Lanza G, Scarpa A, Vecchione A, Negrini M, Harris CC, Croce CM. A microRNA expression signature of human solid tumors defines cancer gene targets. *Proc. Natl. Acad. Sci. U S A.* 2006 Feb 14;103(7):2257-61.

Wang M, Tan LP, Dijkstra MK, van Lom K, Robertus JL, Harms G, Blokzijl T, Kooistra K, van T'veer MB, Rosati S, Visser L, Jongen-Lavrencic M, Kluin PM, van den Berg A. miRNA analysis in B-cell chronic lymphocytic leukaemia: proliferation centres characterized by low miR-150 and high BIC/miR-155 expression. *J. Pathol.* 2008 May;215(1):13-20.

Wang Y, Rathinam R, Walch A, Alahari SK. ST14 (Suppression of Tumorigenicity 14) Gene Is a Target for miR-27b, and the Inhibitory Effect of ST14 on Cell Growth Is Independent of miR-27b Regulation. *J. Biol. Chem.* 2009a Aug 21;284(34):23094-106.

Wang X, Tang S, Le SY, Lu R, Rader JS, Meyers C, Zheng ZM. Aberrant expression of oncogenic and tumor-suppressive microRNAs in cervical cancer is required for cancer cell growth. PLoS ONE. 2008b Jul 2;3(7):e2557.

Wang S, Bian C, Yang Z, Bo Y, Li J, Zeng L, Zhou H, Zhao RC. miR-145 inhibits breast cancer cell growth through RTKN. *Int. J. Oncol.* 2009c May;34(5):1461-6.

Wang Y, Lee CG. MicroRNA and cancer - focus on apoptosis. *J. Cell Mol. Med.* 2009 Jan;13(1):12-23.

Webster RJ, Giles KM, Price KJ, Zhang PM, Mattick JS, Leedman PJ. Regulation of epidermal growth factor receptor signaling in human cancer cells by microRNA-7. *J. Biol. Chem.* 2009 Feb 27;284(9):5731-41.

Weigelt B, Horlings HM, Kreike B, Hayes MM, Hauptmann M, Wessels LF, de Jong D, Van de Vijver MJ, Van't Veer LJ, Peterse JL. Refinement of breast cancer classification by molecular characterization of histological special types. *J. Pathol.* 2008 Oct;216(2):141-50.

Wickramasinghe NS, Manavalan TT, Dougherty SM, Riggs KA, Li Y, Klinge CM. Estradiol downregulates miR-21 expression and increases miR-21 target gene expression in MCF-7 breast cancer cells. *Nucleic Acids Res.* 2009 May;37(8):2584-95.

Wilfred BR, Wang WX, Nelson PT. Energizing miRNA research: a review of the role of miRNAs in lipid metabolism, with a prediction that miR-

103/107 regulates human metabolic pathways. *Mol. Genet. Metab.* 2007 Jul;91(3):209-17.

Wong QW, Lung RW, Law PT, Lai PB, Chan KY, To KF, Wong N. MicroRNA-223 is commonly repressed in hepatocellular carcinoma and potentiates expression of Stathmin1. *Gastroenterology.* 2008 Jul;135(1):257-69.

Woods K, Thomson JM, Hammond SM. Direct regulation of an oncogenic micro-RNA cluster by E2F transcription factors. *J. Biol. Chem.* 2007 Jan 26;282(4):2130-4.

Wu H, Mo YY. Targeting miR-205 in breast cancer. *Expert Opin Ther Targets.* 2009 Dec;13(12):1439-48.

Wu H, Zhu S, Mo YY. Suppression of cell growth and invasion by miR-205 in breast cancer. *Cell Res.* 2009 Apr;19(4):439-48.

Xia L, Zhang D, Du R, Pan Y, Zhao L, Sun S, Hong L, Liu J, Fan D. miR-15b and miR-16 modulate multidrug resistance by targeting BCL2 in human gastric cancer cells. *Int. J .Cancer.* 2008 Jul 15;123(2):372-9.

Yan LX, Huang XF, Shao Q, Huang MY, Deng L, Wu QL, Zeng YX, Shao JY. MicroRNA miR-21 overexpression in human breast cancer is associated with advanced clinical stage, lymph node metastasis and patient poor prognosis.*RNA.* 2008 Nov;14(11):2348-60.

Yanaihara N, Caplen N, Bowman E, Seike M, Kumamoto K, Yi M, Stephens RM, Okamoto A, Yokota J, Tanaka T, Calin GA, Liu CG, Croce CM, Harris CC. Unique microRNA molecular profiles in lung cancer diagnosis and prognosis. *Cancer Cell.* 2006 Mar;9(3):189-98.

Yu F, Yao H, Zhu P, Zhang X, Pan Q, Gong C, Huang Y, Hu X, Su F, Lieberman J, Song E. let-7 regulates self renewal and tumorigenicity of breast cancer cells. *Cell.* 2007 Dec 14;131(6):1109-23.

Yu Z, Wang C, Wang M, Li Z, Casimiro MC, Liu M, Wu K, Whittle J, Ju X, Hyslop T, McCue P, Pestell RG. A cyclin D1/microRNA 17/20 regulatory feedback loop in control of breast cancer cell proliferation. *J. Cell Biol.* 2008 Aug 11;182(3):509-17.

Zhang J, Du YY, Lin YF, Chen YT, Yang L, Wang HJ, Ma D. The cell growth suppressor, mir-126, targets IRS-1. *Biochem. Biophys. Res. Commun.* 2008 Dec 5;377(1):136-40.

Zhang Z, Sun H, Dai H, Walsh RM, Imakura M, Schelter J, Burchard J, Dai X, Chang AN, Diaz RL, Marszalek JR, Bartz SR, Carleton M, Cleary MA, Linsley PS, Grandori C. MicroRNA miR-210 modulates cellular response to hypoxia through the MYC antagonist MNT. *Cell Cycle.* 2009a 2009 Sep 1;8(17):2756-68.

Zhao JJ, Lin J, Yang H, Kong W, He L, Ma X, Coppola D, Cheng JQ. MicroRNA-221/222 negatively regulates estrogen receptor alpha and is associated with tamoxifen resistance in breast cancer. *J. Biol. Chem.* 2008 Nov 7;283(45):31079-86.

Zhu S, Si ML, Wu H, Mo YY. MicroRNA-21 targets the tumor suppressor gene tropomyosin 1 (TPM1). *J. Biol .Chem.* 2007 May 11;282(19):14328-36.

Zhu S, Wu H, Wu F, Nie D, Sheng S, Mo YY. MicroRNA-21 targets tumor suppressor genes in invasion and metastasis. *Cell Res.* 2008 Mar;18(3): 350-9.

Chapter 5

CLINICAL POTENTIAL OF MIRNAS IN BREAST CANCER

ABSTRACT

The association of multiple miRNAs with breast cancer, and the fact that most of these miRNAs may modulate complex functional networks of mRNAs, identify them as potential diagnostic, prognostic and predictive tumor markers, as well as possible therapeutical targets. An additional advantage of miRNAs in oncology is that they are remarkably stable, being notably detectable in serum and plasma.

The data presented in chapter 4 show that the expression level of a number of miRNAs is modified in breast cancer. Some of them clearly appear to function as oncogenes or tumor suppressors. The expression pattern of many miRNAs is in correlation with clinicopathological characteristics of the tumors and clinical outcome (metastasis, for instance).

Currently, the routine assessment of breast cancer is based on two biomarkers: ER-α and HER2/neu [Lacroix *et al.* 2001; Lacroix *et al.* 2004; Ferretti *et al.* 2007]. The longest established breast cancer molecular indicator, the ER-α has been quantified for more than 30 years in tumor samples. This has led to its definitive acceptance both as a prognostic indicator (its expression is associated with longer survival) and a predictor of patient responsiveness to antiestrogens, particularly in lymph node-negative patients. The prognostic relevance of HER2/neu has been demonstrated by numerous studies. Moreover, HER2/neu is a target for antibody-based therapeutic strategies using trastuzumab and its expression level is predictor of patients' response to trastuzumab.

There is a growing need for additional reliable molecular markers, since the perfect marker for breast cancer may not even exist. As a matter of fact, the ideal marker should be produced solely by cancer cells or in their immediate vicinity; it should be specific and sensitive, and easily measurable in a reproducible way through simple, fast, and inexpensive techniques; it should allow estimation of the tumor volume and assessment of the efficacy of therapy and might itself constitute a highly tumor-specific therapeutic target. In reality, the highly heterogeneous nature of breast tumors makes their exhaustive description based on the expression levels of only a few genes impossible. This is further hampered by the diversity of processes (proliferation, adhesion, proteolysis, chemoresistance, hormone sensitivity) that characterize tumor behavior. Accounting for the complexity of tumors undoubtedly requires recourse to a panel of selected indicators [Lacroix *et al.* 2002].

This need for new markers led to the recent introduction of massive messenger RNA (mRNA) analysis by microarrays, resulting in the obtainment of breast cancer-related gene expression signatures and the definition of new subtypes. Indeed, "molecular profiling" is expected to: 1) help to identify and characterize tumors (mRNAs as diagnostic markers, or indicators); 2) allow foreseeing the evolution and the complications - notably metastasis - of tumors (mRNAs as prognostic markers); 3) provide an estimation of the patient's responsiveness to specific therapy (mRNAs as predictive markers). Besides their value as clinical indicators, some of these mRNAs could also be used as therapeutic targets [Peppercorn *et al.* 2008; Cianfrocca and Gradishar 2009].

It is likely that, in a near future, miRNA signatures, which are currently showing capability of accurately classifying tumors according to currently available prognostic variables, will serve as new biomarkers and prognostic indicators. It is even thought that miRNAs have the potential to improve greatly the precision of the recently derived genomic signatures, given that miRNA profiles have superior accuracy to mRNA profiling in this regard [Lu *et al.* 2005]. While breast cancer subtypes have been defined by mRNA expression profiling, a comprehensive interrogation of the breast cancer subclasses via miRNA expression profiling could further characterize the molecular basis underlying these subtypes, perhaps define more precise subsets of breast cancer, and provide opportunities for the identification of novel targets that can be exploited for targeted therapy [Heneghan *et al.* 2009].

One aspect of miRNA biogenesis that makes them particularly attractive as a biomarker is the fact that they are remarkably stable and well preserved in formalin fixed, paraffin embedded tissues as well as fresh snap frozen specimens [Li et al. 2007; Xi et al. 2007; Hasemeier et al. 2008; Hui et al. 2009]. Moreover, they are maintained in a stable state in serum and plasma, protected from endogenous RNase activity [Mitchell et al. 2008; Chim et al. 2008; Chin and Slack 2008], thus allowing the detection of miRNA expression patterns directly from serum, without resorting to an invasive procedure, such as biopsy. In a recent study, miRNA expression patterns were evaluated in human serum for five types of human cancer, prostate, colon, ovarian, breast and lung, using a pan-human miRNA, high density microarray. It was shown that sufficient miRNAs were present in one milliliter of serum to detect miRNA expression patterns, without the need for amplification techniques. In addition, these expression patterns could be used to correctly discriminate between normal and cancer patient samples [Lodes et al. 2009]. In another study, miRNA species could be detected in archived serum from breast cancer patients; this allowed finding that miR-155 may be differentially expressed in the serum of women with progesterone receptor positive tumors compared to women with PR breast cancer [Zhu et al. 2009]. The fact that serum and plasma analysis could accurately detect cancer without resorting to an invasive procedure makes miRNAs also suitable for cancer screening in high-risk populations.

As emerging evidences highlight the importance of miRNAs in diagnosis and prognosis of breast cancer, the usefulness of miRNA-based breast cancer therapy is now being explored. miRNA-based therapeutic strategies seem to offer an alternative for targeting multiple gene networks that are controlled by a single miRNA. These approaches can be formulated by either antagonizing or restoring the functions of miRNAs [Marquez and McCaffrey 2008].

In breast cancer, target miRNAs can be putative oncogenes or tumor suppressor genes also found in various other cancers, such as miR-21, miR-155 (oncogenes), miR-126, miR-145, and let-7 family members (tumor suppressors). Most targets of these miRNAs appear to be linked to cell proliferation and migration. For instance, miR-21 [Iorio et al. 2005] is overexpressed in at least six tumor types (lung, breast, stomach, prostate, colon, and pancreatic tumors) [Volinia et al. 2006] and, according to another other study, the most significantly up-regulated miRNA in breast cancer compared to normal adjacent tumor tissues [Yan et al. 2008]. Anti-miRNA 2-O-methyl or locked nucleic acid oligonucleotides used to inactivate oncomirs such as miR-21 in breast tumors may taper tumor growth. Anti-miR-21-

induced reduction in tumor growth, interestingly, was also shown to be potentiated by the addition of the chemotherapeutic agent topotecan, an inhibitor of DNA topoisomerase I [Si et al. 2007]. This suggests that suppression of the oncogenic miR-21 could sensitize tumor cells to anticancer therapy, which is an exciting prospect for patients exhibiting a poor response to primary chemotherapy [Heneghan et al. 2009].

Conversely, the induction of tumor suppressor miRNA expression using viral or liposomal delivery of tissue-specific tumor suppressors to affected tissue may result in the prevention of progression, or even shrinking, of breast tumors. For instance, miR-145 appears as one of the most consistently down-regulated miRNAs in breast cancer compared to normal tissue, and early manifestation of altered miR-145 expression may be detected in atypical hyperplasia and carcinoma in situ lesions. In addition, miR-145 may be detected in serum samples from cancer and normal patients [Zhu et al. 2009]. miR-145 re-expression in breast cancer was accompanied by a pro-apoptotic effect, dependent on TP53 activation, while TP53 activation can, in turn, stimulate miR-145 expression, thus establishing a death-promoting loop between miR-145 and TP53. Moreover, miR-145 can downregulate ER-α protein expression through direct interaction with two complementary sites within its coding sequence. This supports a view that miR-145 re-expression therapy could be mainly envisioned in the specific group of patients with ER-α -positive and/or TP53 wild-type tumors [Spizzo et al. 2010].

Other target miRNAs may be associated to more specific pathways, such as the HER2/neu family-mediated or ER-α-driven signaling. For instance, miR-18a, miR-22, miR-181, miR-206, miR-221/222 have been shown to be involved in down-regulation of ER-α expression and in suppression of ER-α-mediated signaling in BCC see chapter 4). By regulating the signatures of these miRNAs, it is likely to find a better way for treating ER-α-negative breast cancer. miR-205 is able to interfere with the proliferative pathway mediated by ERBB receptor family (including HER2/neu), and to increase the responsiveness of BCC to tyrosine-kinase inhibitors gefitinib and lapatinib. This suggests that restoration of miR-205 could improve the responsiveness of BCC to specific anticancer therapies [Iorio et al. 2009]. Infection of BCC with retroviral constructs expressing either miR-125a or miR-125b resulted in suppression of HER2/neu [Scott et al. 2007]. Such miRNA could be another target to investigate the mechanism of action of HER2/neu inhibitors (such as trastuzumab) [Chen and Russo 2009]

Other potential target miRNAs are miR-210, miR-451 and miR-34a. miR-210 is associated to processes triggered by hypoxia, which may lead to increased angiogenesis in cancer [Ivan *et al.* 2008]. A study showed that doxorubicin (DOX)-resistant BCC exhibited alterations in miRNA profile and miR-451 was identified to regulate the expression of multidrug resistance 1 gene [Kovalchuk *et al.* 2008]. Further investigation showed that restoring miR-451 into DOX-resistant BCC resulted in the increased sensitivity of cells to DOX, indicating that restoration of such altered miRNA expression may have important implications for breast cancer therapy. Other miRNAs that seem involved in drug resistance and could constitute clinical targets are miR-326 [Liang *et al.* 2009] and miR-328 [Pan *et al.* 2009]. miR-34a is a direct target of p53 [Lodygin *et al.* 2008]. It would be interesting to investigate if miR-34a is deregulated in breast cancer patients carrying wild-type p53. If so, this would be one of the mechanisms used by tumor cells to escape the apoptotic control by p53 and to survive under oncogenic circumstance. Thus, miRNA-based therapy by restoring miR-34a function in breast cancer could also be a strategy to improve responsiveness to chemotherapy.

Of peculiar interest are miR-200 family members and miR-205, which appear to exert a significant role in the control of epithelial-mesenchymal transition (EMT) [Korpal and Kang 2008]. EMT is often presented as an essential step in order for BCC to escape from the primary tumor. miRNA-based approaches could help to counteract the metastatic process which is still incurable in breast cancer.

REFERENCES

Chen JQ, Russo J. ERalpha-negative and triple negative breast cancer: Molecular features and potential therapeutic approaches. *Biochim. Biophys. Acta.* 2009 Dec;1796(2):162-75.

Chim SS, Shing TK, Hung EC, Leung TY, Lau TK, Chiu RW, Lo YM. Detection and characterization of placental microRNAs in maternal plasma. *Clin. Chem.* 2008 Mar;54(3):482-90.

Chin LJ, Slack FJ. A truth serum for cancer--microRNAs have major potential as cancer biomarkers. *Cell Res.* 2008 Oct;18(10):983-4.

Cianfrocca M, Gradishar W. New molecular classifications of breast cancer. CA *Cancer J. Clin.* 2009 Sep-Oct;59(5):303-13.

Ferretti G, Felici A, Papaldo P, Fabi A, Cognetti F. HER2/neu role in breast cancer: from a prognostic foe to a predictive friend. *Curr. Opin. Obstet. Gynecol.* 2007 Feb;19(1):56-62.

Hasemeier B, Christgen M, Kreipe H, Lehmann U. Reliable microRNA profiling in routinely processed formalin-fixed paraffin-embedded breast cancer specimens using fluorescence labelled bead technology. *BMC Biotechnol.* 2008 Nov 27;8:90.

Heneghan HM, Miller N, Lowery AJ, Sweeney KJ, Kerin MJ. MicroRNAs as Novel Biomarkers for Breast Cancer. *J Oncol.* 2009;2009:950201.

Hui AB, Shi W, Boutros PC, Miller N, Pintilie M, Fyles T, McCready D, Wong D, Gerster K, Waldron L, Jurisica I, Penn LZ, Liu FF. Robust global micro-RNA profiling with formalin-fixed paraffin-embedded breast cancer tissues. *Lab. Invest.* 2009 May;89(5):597-606.

Iorio MV, Ferracin M, Liu CG, Veronese A, Spizzo R, Sabbioni S, Magri E, Pedriali M, Fabbri M, Campiglio M, Ménard S, Palazzo JP, Rosenberg A, Musiani P, Volinia S, Nenci I, Calin GA, Querzoli P, Negrini M, Croce CM. MicroRNA gene expression deregulation in human breast cancer. *Cancer Res.* 2005 Aug 15;65(16):7065-70.

Ivan M, Harris AL, Martelli F, Kulshreshtha R. Hypoxia response and microRNAs: no longer two separate worlds. *J. Cell Mol. Med.* 2008 Sep-Oct;12(5A):1426-31.

Korpal M, Kang Y. The emerging role of miR-200 family of microRNAs in epithelial-mesenchymal transition and cancer metastasis. *RNA Biol.* 2008 Jul-Sep;5(3):115-9.

Kovalchuk O, Filkowski J, Meservy J, Ilnytskyy Y, Tryndyak VP, Chekhun VF, Pogribny IP. Involvement of microRNA-451 in resistance of the MCF-7 breast cancer cells to chemotherapeutic drug doxorubicin. *Mol. Cancer Ther.* 2008 Jul;7(7):2152-9.

Lacroix M, Querton G, Hennebert P, Larsimont D, Leclercq G. Estrogen receptor analysis in primary breast tumors by ligand-binding assay, immunocytochemical assay, and northern blot: a comparison. *Breast Cancer Res. Treat.* 2001 Jun;67(3):263-71.

Lacroix M, Zammatteo N, Remacle J, Leclercq G. A low-density DNA microarray for analysis of markers in breast cancer. *Int. J. Biol. Markers.* 2002 Jan-Mar;17(1):5-23.

Lacroix M, Toillon RA, Leclercq G. Stable 'portrait' of breast tumors during progression: data from biology, pathology and genetics. *Endocr. Relat. Cancer.* 2004 Sep;11(3):497-522.

Li J, Smyth P, Flavin R, Cahill S, Denning K, Aherne S, Guenther SM, O'Leary JJ, Sheils O.

Comparison of miRNA expression patterns using total RNA extracted from matched samples of formalin-fixed paraffin-embedded (FFPE) cells and snap frozen cells. *BMC Biotechnol.* 2007 Jun 29;7:36.

Liang Z, Wu H, Xia J, Li Y, Zhang Y, Huang K, Wagar N, Yoon Y, Cho HT, Scala S, Shim H. Involvement of miR-326 in chemotherapy resistance of breast cancer through modulating expression of multidrug resistance-associated protein 1. *Biochem Pharmacol.* 2010 Mar 15;79(6):817-24.

Lodes MJ, Caraballo M, Suciu D, Munro S, Kumar A, Anderson B. Detection of cancer with serum miRNAs on an oligonucleotide microarray. PLoS One. 2009 Jul 14;4(7):e6229.

Lodygin D, Tarasov V, Epanchintsev A, Berking C, Knyazeva T, Körner H, Knyazev P, Diebold J, Hermeking H. Inactivation of miR-34a by aberrant CpG methylation in multiple types of cancer. *Cell Cycle.* 2008 Aug 15;7(16):2591-600.

Lu J, Getz G, Miska EA, Alvarez-Saavedra E, Lamb J, Peck D, Sweet-Cordero A, Ebert BL, Mak RH, Ferrando AA, Downing JR, Jacks T, Horvitz HR, Golub TR. MicroRNA expression profiles classify human cancers. *Nature.* 2005 Jun 9;435(7043):834-8.

Marquez RT, McCaffrey AP. Advances in microRNAs: implications for gene therapists. *Hum. Gene Ther.* 2008 Jan;19(1):27-38.

Mitchell PS, Parkin RK, Kroh EM, Fritz BR, Wyman SK, Pogosova-Agadjanyan EL, Peterson A, Noteboom J, O'Briant KC, Allen A, Lin DW, Urban N, Drescher CW, Knudsen BS, Stirewalt DL, Gentleman R, Vessella RL, Nelson PS, Martin DB, Tewari M. Circulating microRNAs as stable blood-based markers for cancer detection. *Proc. Natl. Acad. Sci. U S A.* 2008 Jul 29;105(30):10513-8.

Ng EK, Wong CL, Ma ES, Kwong A. MicroRNAs as New Players for Diagnosis, Prognosis, and Therapeutic Targets in Breast Cancer. *J. Oncol.* 2009;2009:305420. Epub 2009 Jul 27.

Pan YZ, Morris ME, Yu AM. MicroRNA-328 negatively regulates the expression of breast cancer resistance protein (BCRP/ABCG2) in human cancer cells. *Mol. Pharmacol.* 2009 Jun;75(6):1374-9.

Peppercorn J, Perou CM, Carey LA. Molecular subtypes in breast cancer evaluation and management: divide and conquer. *Cancer Invest.* 2008 Feb;26(1):1-10.

Scott GK, Goga A, Bhaumik D, Berger CE, Sullivan CS, Benz CC. Coordinate suppression of ERBB2 and ERBB3 by enforced expression of micro-RNA miR-125a or miR-125b. *J. Biol. Chem.* 2007 Jan 12;282(2):1479-86.

Si ML, Zhu S, Wu H, Lu Z, Wu F, Mo YY. miR-21-mediated tumor growth. *Oncogene.* 2007 Apr 26;26(19):2799-803.

Spizzo R, Nicoloso MS, Lupini L, Lu Y, Fogarty J, Rossi S, Zagatti B, Fabbri M, Veronese A, Liu X, Davuluri R, Croce CM, Mills G, Negrini M, Calin GA. miR-145 participates with TP53 in a death-promoting regulatory loop and targets estrogen receptor-alpha in human breast cancer cells. *Cell Death Differ.* 2010 Feb;17(2):246-54.

Volinia S, Calin GA, Liu CG, Ambs S, Cimmino A, Petrocca F, Visone R, Iorio M, Roldo C, Ferracin M, Prueitt RL, Yanaihara N, Lanza G, Scarpa A, Vecchione A, Negrini M, Harris CC, Croce CM. A microRNA expression signature of human solid tumors defines cancer gene targets. *Proc Natl Acad Sci U S A.* 2006 Feb 14;103(7):2257-61.

Xi Y, Nakajima G, Gavin E, Morris CG, Kudo K, Hayashi K, Ju J. Systematic analysis of microRNA expression of RNA extracted from fresh frozen and formalin-fixed paraffin-embedded samples. *RNA.* 2007 Oct;13(10):1668-74.

Yan LX, Huang XF, Shao Q, Huang MY, Deng L, Wu QL, Zeng YX, Shao JY. MicroRNA miR-21 overexpression in human breast cancer is associated with advanced clinical stage, lymph node metastasis and patient poor prognosis.*RNA.* 2008 Nov;14(11):2348-60.

Zhu W, Qin W, Atasoy U, Sauter ER. Circulating microRNAs in breast cancer and healthy subjects. *BMC Res. Notes.* 2009 May 19;2:89.

Chapter 6

HUMAN MIRNAS: GENES, NAMES, LOCI, SEQUENCES, CLUSTERS

Table 1. List of human miRNA identified to date, with their specificities

miRNA gene identifier	miRNA gene locus	miRNA precursor (stem-loop) identifier	mature miRNA identifier	mature miRNA sequence
MIRLET7A1	9q22.32	let-7a-1	let-7a	ugagguaguagguuguauaguu
MIRLET7A2	11q24.1	let-7a-2	let-7a	ugagguaguagguuguauaguu
MIRLET7A3	22q13.31	let-7a-3	let-7a	ugagguaguagguuguauaguu
MIRLET7B	22q13.31	let-7b	let-7b	ugagguaguagguugugugguu
MIRLET7C	21q21.1	let-7c	let-7c	ugagguaguagguuguaugguu
MIRLET7D	9q22.32	let-7d	let-7d	agagguaguagguugcauaguu
MIRLET7E	19q13.33	let-7e	let-7e	ugagguaggagguuguauaguu
MIRLET7F1	9q22.32	let-7f-1	let-7f	ugagguaguagauuguauaguu
MIRLET7F2	Xp11.22	let-7f-2	let-7f	ugagguaguagauuguauaguu
MIRLET7G	3p21.1	let-7g	let-7g	ugagguaguaguuuguacaguu
MIRLET7I	12q14.1	let-7i	let-7i	ugagguaguaguuugugcuguu
MIR1-1	20q13.33	mir-1-1	miR-1	uggaauguaaagaaguauguau
MIR1-2	18q11.2	mir-1-2	miR-1	uggaauguaaagaaguauguau
MIR7-1	9q21.32	mir-7-1	miR-7	uggaagacuagugauuuuguugu
MIR7-2	15q26.1	mir-7-2	miR-7	uggaagacuagugauuuuguugu
MIR7-3	19p13.3	mir-7-3	miR-7	uggaagacuagugauuuuguugu
MIR9-1	1q22	mir-9-1	miR-9	ucuuugguuaucuagcuguauga
MIR9-2	5q14.3	mir-9-2	miR-9	ucuuugguuaucuagcuguauga
MIR9-3	15q26.1	mir-9-3	miR-9	ucuuugguuaucuagcuguauga
MIR10A	17q21.32	mir-10a	miR-10a	uacccuguagauccgaauuugug

miRNA gene identifier	miRNA gene locus	miRNA precursor (stem-loop) identifier	mature miRNA identifier	mature miRNA sequence
MIR10B	2q31.1	mir-10b	miR-10b	uacccuguagaaccgaauuugug
MIR15A	13q14.3	mir-15a	miR-15a	uagcagcacauaauggguuugug
MIR15B	3q25.33	mir-15b	miR-15b	uagcagcacaucaugguuuaca
MIR16-1	13q14.3	mir-16-1	miR-16	uagcagcacguaaauauuggcg
MIR16-2	3q25.33	mir-16-2	miR-16	uagcagcacguaaauauuggcg
MIR17	13q31.3	mir-17	miR-17	caaagugcuuacagugcagguag
MIR18A	13q31.3	mir-18a	miR-18a	uaaggugcaucuagugcagauag
MIR18B	Xq26.2	mir-18b	miR-18b	uaaggugcaucuagugcaguag
MIR19A	13q31.3	mir-19a	miR-19a	ugugcaaaucuaugcaaaacuga
MIR19B-1	13q31.3	mir-19b-1	miR-19b	ugugcaaauccaugcaaaacuga
MIR19B-2	Xq26.2	mir-19b-2	miR-19b	ugugcaaauccaugcaaaacuga
MIR20A	13q31.3	mir-20a	miR-20a	uaaagugcuuauagugcagguag
MIR20B	Xq26.2	mir-20b	miR-20b	caaagugcucauagugcagguag
MIR21	17q23.1	mir-21	miR-21	uagcuuaucagacugauguuga
MIR22	17p13.3	mir-22	miR-22	aagcugccaguugaagaacugu
MIR23A	19p13.12	mir-23a	miR-23a	aucacauugccagggauuucc
MIR23B	9q22.32	mir-23b	miR-23b	aucacauugccagggauuacc
MIR24-1	9q22.32	mir-24-1	miR-24	uggcucaguucagcaggaacag
MIR24-2	19p13.12	mir-24-2	miR-24	uggcucaguucagcaggaacag
MIR25	7q22.1	mir-25	miR-25	cauugcacuugucucggucuga
MIR26A-1	3p22.2	mir-26a-1	miR-26a	uucaaguaauccaggauaggcu
MIR26A-2	12q14.1	mir-26a-2	miR-26a	uucaaguaauccaggauaggcu
MIR26B	2q35	mir-26b	miR-26b	uucaaguaauucaggauaggu
MIR27A	19p13.12	mir-27a	miR-27a	uucacaguggcuaaguuccgc
MIR27B	9q22.32	mir-27b	miR-27b	uucacaguggcuaaguucugc
MIR28	3q28	mir-28	miR-28-5p	aaggagcucacagucuauugag
			miR-28-3p	cacuagauugugagcuccugga
MIR29A	7q32.3	mir-29a	miR-29a	uagcaccaucugaaaucgguua
MIR29B-1	7q32.3	mir-29b-1	miR-29b	uagcaccauuugaaaucaguguu
MIR29B-2	1q32.2	mir-29b-2	miR-29b	uagcaccauuugaaaucaguguu
MIR29C	1q32.2	mir-29c	miR-29c	uagcaccauuugaaaucgguua
MIR30A	6q13	mir-30a	miR-30a	uguaaacauccucgacuggaag
MIR30B	8q24.22	mir-30b	miR-30b	uguaaacauccuacacucagcu
MIR30C1	1p34.2	mir-30c-1	miR-30c	uguaaacauccuacacucucagc
MIR30C2	6q13	mir-30c-2	miR-30c	uguaaacauccuacacucucagc
MIR30D	8q24.22	mir-30d	miR-30d	uguaaacauccccgacuggaag
MIR30E	1p34.2	mir-30e	miR-30e	uguaaacauccuugacuggaag
MIR31	9p21.3	mir-31	miR-31	aggcaagaugcuggcauagcu
MIR32	9q31.3	mir-32	miR-32	uauugcacauuacuaaguugca
MIR33A	22q13.2	mir-33a	miR-33a	gugcauuguaguugcauugca
MIR33B	17p11.2	mir-33b	miR-33b	gugcauugcuguugcauugc
MIR34A	1p36.22	mir-34a	miR-34a	uggcagugucuuagcugguugu
MIR34B	11q23.1	mir-34b	miR-34b	caaucacuaaccucacugccau
MIR34C	11q23.1	mir-34c	miR-34c-5p	aggcaguguaguuagcugauugc
			miR-34c-3p	aaucacuaaccacacggccagg

miRNA gene identifier	miRNA gene locus	miRNA precursor (stem-loop) identifier	mature miRNA identifier	mature miRNA sequence
MIR92A1	13q31.3	mir-92a-1	miR-92a	uauugcacuugucccggccugu
MIR92A2	Xq26.2	mir-92a-2	miR-92a	uauugcacuugucccggccugu
MIR92B	1q22	mir-92b	miR-92b	uauugcacucgucccggccucc
MIR93	7q22.1	mir-93	miR-93	caaagugcuguucgugcagguag
MIR95	4p16.1	mir-95	miR-95	uucaacggguauuuauugagca
MIR96	7q32.2	mir-96	miR-96	uuuggcacuagcacauuuuugcu
MIR98	Xp11.22	mir-98	miR-98	ugagguaguaaguuguauuguu
MIR99A	21q21.1	mir-99a	miR-99a	aacccguagauccgaucuugug
MIR99B	19q13.33	mir-99b	miR-99b	cacccguagaaccgaccuugcg
MIR100	11q24.1	mir-100	miR-100	aacccguagauccgaacuugug
MIR101-1	1p31.3	mir-101-1	miR-101	uacaguacugugauaacugaa
MIR101-2	9p24.1	mir-101-2	miR-101	uacaguacugugauaacugaa
MIR103-1	5q35.1	mir-103-1	miR-103	agcagcauuguacagggcuauga
MIR103-2	20p13	mir-103-2	miR-103	agcagcauuguacagggcuauga
MIR103-1AS	5q34	mir-103-1-as	miR-103-as	ucauagcccuguacaaugcugcu
MIR103-2AS	20p13	mir-103-1-as	miR-103-as	ucauagcccuguacaaugcugcu
MIR105-1	Xq28	mir-105-1	miR-105	ucaaaugcucagacuccuguggu
MIR105-2	Xq28	mir-105-2	miR-105	Ucaaaugcucagacuccuguggu
MIR106A	Xq26.2	mir-106a	miR-106a	aaaagugcuuacagugcagguag
MIR106B	7q22.1	mir-106b	miR-106b	uaaagugcugacagugcagau
MIR107	10q23.31	mir-107	miR-107	agcagcauuguacagggcuauca
MIR122	18q21.31	mir-122	miR-122	uggagugugacaauggguuug
MIR124-1	8p23.1	mir-124-1	miR-124	uaaggcacgcggugaaugcc
MIR124-2	8q12.3	mir-124-2	miR-124	uaaggcacgcggugaaugcc
MIR124-3	20q13.33	mir-124-3	miR-124	uaaggcacgcggugaaugcc
MIR125A	19q13.33	mir-125a	miR-125a-5p	ucccugagaccccuuuaaccuguga
			miR-125a-3p	acaggugagguucuugggagcc
MIR125B1	11q24.1	mir-125b-1	miR-125b	ucccugagacccuaacuuguga
MIR125B2	21q21.1	mir-125b-2	miR-125b	ucccugagacccuaacuuguga
MIR126	9q34.3	mir-126	miR-126	ucguaccgugaguaauaaugcg
MIR127	14q32.31	mir-127	miR-127-5p	cugaagcucagagggcucugau
			miR-127-3p	ucggauccgucugagcuuggcu
MIR128-1	2q21	mir-128-1	miR-128	ucacagugaaccggucucuuu
MIR128-2	3p22	mir-128-2	miR-128	ucacagugaaccggucucuuu
MIR129-1	7q32.1	mir-129-1	miR-129-5p	cuuuugcggucugggcuugc
			miR-129-3p	aagcccuuaccccaaaaaguau
MIR129-2	11p11.2	mir-129-2	miR-129-5p	cuuuugcggucugggcuugc
			miR-129-3p	aagcccuuaccccaaaaagcau
MIR130A	11q12.1	mir-130a	miR-130a	cagugcaauguuaaagggcau

miRNA gene identifier	miRNA gene locus	miRNA precursor (stem-loop) identifier	mature miRNA identifier	mature miRNA sequence
MIR130B	22q11.21	mir-130b	miR-130b	cagugcaaugaugaaagggcau
MIR132	17p13.3	mir-132	miR-132	uaacagucuacagccauggucg
MIR133A1	18q11.2	mir-133a-1	miR-133a	uuuggucccuucaaccagcug
MIR133A2	20q13.33	mir-133a-2	miR-133a	uuuggucccuucaaccagcug
MIR133B	6p12.2	mir-133b	miR-133b	uuuggucccuucaaccagcua
MIR134	14q32.31	mir-134	miR-134	ugugacugguugaccagaggg
MIR135A1	3p21.1	mir-135a-1	miR-135a	uauggcuuuuauuccuauguga
MIR135A2	12q23.1	mir-135a-2	miR-135a	uauggcuuuuauuccuauguga
MIR135B	1q32.1	mir-135b	miR-135b	uauggcuuucauuccuauguga
MIR136	14q32.31	mir-136	miR-136	acuccauuuguuuugaugauga
MIR137	1p21.3	mir-137	miR-137	uuauugcuuaagaauacgcguag
MIR138-1	3p21.32	mir-138-1	miR-138	agcugguguugaaucaggccg
MIR138-2	16q13	mir-138-2	miR-138	agcugguguugaaucaggccg
MIR139	11q13.4	mir-139	miR-139-5p	ucuacagugcacgugucuccag
			miR-139-3p	ggagacgcggcccuguuggagu
MIR140	16q22.1	mir-140	miR-140-5p	cagugguuuuacccuauggua g
			miR-140-3p	uaccacagggu agaaccacgg
MIR141	12p13.31	mir-141	miR-141	uaacacugucugguaaagaugg
MIR142	17q22	mir-142	miR-142-5p	cauaaaguagaaagcacuacu
			miR-142-3p	uguaguguuccuacuuuaugga
MIR143	5q32	mir-143	miR-143	ugagaugaagcacuguagcuc
MIR144	17q11.2	mir-144	miR-144	uacaguauagaugauguacu
MIR145	5q32	mir-145	miR-145	guccaguuucccaggaaucccu
MIR146A	5q33.3	mir-146a	miR-146a	ugagaacugaauuccauggguu
MIR146B	10q24.32	mir-146b	miR-146b-5p	ugagaacugaauuccauaggcu
			miR-146b-3p	ugcccuguggacucaguucugg
MIR147	9q33.2	mir-147	miR-147	guguguggaaaugcuucugc
MIR147B	15q21.1	mir-147b	miR-147b	gugugcggaaaugcuucugcua
MIR148A	7p15.2	mir-148a	miR-148a	ucagugcacuacagaacuuugu
MIR148B	12q13.13	mir-148b	miR-148b	ucagugcaucacagaacuuugu
MIR149	2q37.3	mir-149	miR-149	ucuggcuccugucuucacuccc
MIR150	19q13.33	mir-150	miR-150	ucucccaacccuuguaccagug
MIR151	8q24.3	mir-151	miR-151-5p	ucgaggagcucacagucuagu
			miR-151-3p	cuagacugaagcuccuugagg
MIR152	17q21.32	mir-152	miR-152	ucagugcaugacagaacuugg
MIR153-1	2q35	mir-153-1	miR-153	uugcauagucacaaaagugauc
MIR153-2	7q36.3	mir-153-2	miR-153	uugcauagucacaaaagugauc
MIR154	14q32.31	mir-154	miR-154	uagguuauccguguugccuucg
MIR155	21q21.2	mir-155	miR-155	uuaaugcuaaucgugauaggggu
MIR181A1	1q31.3	mir-181a-1	miR-181a	aacauucaacgcugucggugagu
MIR181A2	9q33.3	mir-181a-2	miR-181a	aacauucaacgcugucggugagu
MIR181B1	1q31.3	mir-181b-1	miR-181b	aacauucauugcugucggugggu
MIR181B2	9q33.3	mir-181b-2	miR-181b	aacauucauugcugucggugggu
MIR181C	19p13.12	mir-181c	miR-181c	aacauucaaccugucggugagu
MIR181D	19p13.13	mir-181d	miR-181d	aacauucauuguugucggugggu

miRNA gene identifier	miRNA gene locus	miRNA precursor (stem-loop) identifier	mature miRNA identifier	mature miRNA sequence
MIR182	7q32.2	mir-182	miR-182	uuuggcaaugguagaacucacacu
MIR183	7q32.2	mir-183	miR-183	uauggcacugguagaauucacu
MIR184	15q25.1	mir-184	miR-184	uggacggagaacugauaagggu
MIR185	22q11.21	mir-185	miR-185	uggagagaaaggcaguuccuga
MIR186	1p31.1	mir-186	miR-186	caaagaaucuccuuuugggcu
MIR187	18q12.2	mir-187	miR-187	ucgugucuuguguugcagccgg
MIR188	Xp11.23	mir-188	miR-188-5p	caucccuugcaugguggaggg
			miR-188-3p	cucccacaugcagggguuugca
MIR190	15q22.2	mir-190	miR-190	ugauauguuugauauauuaggu
MIR190B	1q21.3	mir-190b	miR-190b	ugauauguuugauauugggguu
MIR191	3p21.31	mir-191	miR-191	caacggaaucccaaaagcagcug
MIR192	11q13.1	mir-192	miR-192	cugaccuaugaauugacagcc
MIR193A	17q11.2	mir-193a	miR-193a-5p	ugggucuuugcgggcgagauga
			miR-193a-3p	aacuggccuacaaagucccagu
MIR193B	16p13.12	mir-193b	miR-193b	aacugcccucaaagucccgcu
MIR194-1	1q41	mir-194-1	miR-194	uguaacagcaacuccaugugga
MIR194-2	11q13.1	mir-194-2	miR-194	uguaacagcaacuccaugugga
MIR195	17p13.1	mir-195	miR-195	uagcagcacagaaauauuggc
MIR196A1	17q21.32	mir-196a-1	miR-196a	uagguaguuucauguuguuggg
MIR196A2	12q13.13	mir-196a-2	miR-196a	uagguaguuucauguuguuggg
MIR196B	7p15.2	mir-196b	miR-196b	uagguaguuucuguuguuggg
MIR197	1p13.3	mir-197	miR-197	uucaccaccuucuccacccagc
MIR198	3q13.33	mir-198	miR-198	gguccagagggagauagguuc
MIR199A1	19p13.2	mir-199a-1	miR-199a-5p	cccaguguucagacuaccuguuc
			miR-199a-3p	acaguagucugcacauugguua
MIR199A2	1q24.3	mir-199a-2	miR-199a-5p	cccaguguucagacuaccuguuc
			miR-199a-3p	acaguagucugcacauugguua
MIR199B	9q34.11	mir-199b	miR-199b-5p	cccaguguuuagacuaucuguuc
			miR-199b-3p	acaguagucugcacauugguua
MIR200A	1p36.33	mir-200a	miR-200a	uaacacugucugguaacgaugu
MIR200B	1p36.33	mir-200b	miR-200b	uaauacugccugguaaugauga
MIR200C	12p13.31	mir-200c	miR-200c	uaauacugccggguaaugauga
MIR202	10q26.3	mir-202	miR-202	agagguaagggcaugggaa
MIR203	14q32.33	mir-203	miR-203	gugaaauguuuaggaccacuag
MIR204	9q21.12	mir-204	miR-204	uucccuuugucauccuaugccu
MIR205	1q32.2	mir-205	miR-205	uccuucauuccaccggagucug
MIR206	6p12.2	mir-206	miR-206	uggaauguaaggaagugugugg
MIR208A	14q11.2	mir-208a	miR-208a	auaagacgagcaaaaagcuugu
MIR208B	14q11.2	mir-208b	miR-208b	auaagacgaacaaaagguuugu
MIR210	11p15.5	mir-210	miR-210	cugugcgugugacagcggcuga
MIR211	15q13.3	mir-211	miR-211	uucccuuugucauccuucgccu
MIR212	17p13.3	mir-212	miR-212	uaacagucuccagucacggcc

miRNA gene identifier	miRNA gene locus	miRNA precursor (stem-loop) identifier	mature miRNA identifier	mature miRNA sequence
MIR214	1q24.3	mir-214	miR-214	acagcaggcacagacaggcagu
MIR215	1q41	mir-215	miR-215	augaccuaugaauugacagac
MIR216A	2p16.1	mir-216a	miR-216a	uaaucucagcuggcaacuguga
MIR216B	2p16.1	mir-216b	miR-216b	aaaucucugcaggcaaauguga
MIR217	2p16.1	mir-217	miR-217	uacugcaucaggaacugauugga
MIR218-1	4p15.31	mir-218-1	miR-218	uugugcuugaucuaaccaugu
MIR218-2	5q34	mir-218-2	miR-218	uugugcuugaucuaaccaugu
MIR219-1	6p21.32	mir-219-1	miR-219-5p	ugauugaccaaacgcaauucu
			miR-219-1-3p	agaguugagucuggacgucccg
MIR219-2	9q34.11	mir-219-2	miR-219-5p	ugauugaccaaacgcaauucu
			miR-219-2-3p	agaaugugggcuggacaucugu
MIR220A	Xq25	mir-220a	miR-220a	ccacaccguaucugacacuuu
MIR220B	19p13.3	mir-220b	miR-220b	ccaccaccgugucugacacuu
MIR220C	19q13.33	mir-220c	miR-220c	acacagggcuguugugaagacu
MIR221	Xp11.3	mir-221	miR-221	agcuacauugucugcugggunuc
MIR222	Xp11.3	mir-222	miR-222	agcuacaucuggcuacugggu
MIR223	Xq12	mir-223	miR-223	ugucaguuugucaaauacccca
MIR224	Xq28	mir-224	miR-224	caagucacuagugguuccguu
MIR296	20q13.32	mir-296	miR-296-5p	agggcccccccucaauccugu
			miR-296-3p	gagggunggguggaggcucucc
MIR297	4q25	mir-297	miR-297	auguaugugugcaugugcaug
MIR298	20q13.32	mir-298	miR-298	agcagaagcagggagguucuccca
MIR299	14q32.31	mir-299	miR-299-5p	ugguuuaccgucccacauacau
			miR-299-3p	uauggggaugguaaaccgcuu
MIR300	14q32.31	mir-300	miR-300	uauacaagggcagacucucucu
MIR301A	17q22	mir-301a	miR-301a	cagugcaauaguauugucaaagc
MIR301B	22q11.21	mir-301b	miR-301b	cagugcaaugauauugucaaagc
MIR302A	4q25	mir-302a	miR-302a	uaagugcuuccauguuuugguga
MIR302B	4q25	mir-302b	miR-302b	uaagugcuuccauguuuuaguag
MIR302C	4q25	mir-302c	miR-302c	uaagugcuuccauguuucagugg
MIR302D	4q25	mir-302d	miR-302d	uaagugcuuccauguuugagugu
MIR302E	11p15.4	mir-302e	miR-302e	uaagugcuuccaugcuu
MIR302F	18q12.1	mir-302f	miR-302f	uaauugcuuccauguuu
MIR320A	8p21.3	mir-320a	miR-320a	aaaagcugggguugagagggcga
MIR320B1	1p13.1	mir-320b-1	miR-320b	aaaagcugggguugagagggcaa
MIR320B2	1q42.11	mir-320b-2	miR-320b	aaaagcugggguugagagggcaa
MIR320C1	18q11.2	mir-320c-1	miR-320c	aaaagcugggguugagagggu
MIR320C2	18q11.2	mir-320c-2	miR-320c	aaaagcugggguugagagggu
MIR320D1	13q14.11	mir-320d-1	miR-320d	aaaagcugggguugagagga
MIR320D2	Xq27.1	mir-320d-2	miR-320d	aaaagcugggguugagagga
MIR323	14q32.31	mir-323	miR-323-5p	aggugguccguggcgcguucgc
			miR-323-3p	cacauuacacggucgaccucu
MIR324	17p13.1	mir-324	miR-324-5p	cgcaucccuagggcauuggugu
			miR-324-3p	acugccccaggugcugcugg

miRNA gene identifier	miRNA gene locus	miRNA precursor (stem-loop) identifier	mature miRNA identifier	mature miRNA sequence
MIR325	Xq21.1	mir-325	miR-325	ccuaguagguguccaguaagugu
MIR326	11q13.4	mir-326	miR-326	ccucugggcccuuccuccag
MIR328	16q22.1	mir-328	miR-328	cuggcccucucugcccuuccgu
MIR329-1	14q32.31	mir-329-1	miR-329	aacacaccugguuaaccucuuu
MIR329-2	14q32.31	mir-329-2	miR-329	aacacaccugguuaaccucuuu
MIR330	19q13.32	mir-330	miR-331-5p	ucucugggccugugucuuaggc
			miR-331-3p	gcaaagcacacggccugcagaga
MIR331	12q22	mir-331	miR-331-5p	cuagguaugguccagggaucc
			miR-331-3p	gccccugggccuauccuagaa
MIR335	7q32.2	mir-335	miR-335	ucaagagcaauaacgaaaaaugu
MIR337	14q32.31	mir-337	miR-337-5p	gaacggcuucaucaggaguu
			miR-337-3p	cuccuauaugaugccuuucuuc
MIR338	17q25.3	mir-338	miR-338-5p	aacaauauccuggugcugagug
			miR-338-3p	uccagcaucagugauuuuguug
MIR339	7p22.3	mir-339	miR-339-5p	ucccuguccuccaggagcucacg
			miR-339-3p	ugagcgccucgacgacagagccg
MIR340	5q35.3	mir-340	miR-340	uuauaaagcaaugagacugauu
MIR342	14q32.2	mir-342	miR-342-5p	aggggugcuaucugugauuga
			miR-342-3p	ucucacacagaaaucgcacccgu
MIR345	14q32.2	mir-345	miR-345	gcugacuccuagucccagggcuc
MIR346	10q23.2	mir-346	miR-346	ugucugcccgcaugccugccucu
MIR361	Xq21.2	mir-361	miR-361-5p	uuaucagaaucuccaggggguac
			miR-361-3p	uccccccaggugugauucugauuu
MIR362	Xp11.23	mir-362	miR-362-5p	aauccuuggaaccuaggugugagu
			miR-362-3p	aacacaccuauucaaggauuca
MIR363	Xq26.2	mir-363	miR-363	aauugcacgguauccaucugua
MIR365-1	16p13.12	mir-365-1	miR-365	uaaugccccuaaaaauccuuau
MIR365-2	17q11.2	mir-365-2	miR-365	uaaugccccuaaaaauccuuau
MIR367	4q25	mir-367	miR-367	aauugcacuuuagcaauggugu
MIR369	14q32.31	mir-369	miR-369-5p	agaucgaccguguuauauucgc
			miR-369-3p	aauaauacaugguugaucuuu
MIR370	14q32.31	mir-370	miR-370	gccugcuggggguggaaccuggu
MIR371	19q13.42	mir-371	miR-371-5p	acucaaacuguggggggcacu
			miR-371-3p	aagugccgccaucuuuugagugu
MIR372	19q13.42	mir-372	miR-372	aaagugcugcgacauuugagcgu
MIR373	19q13.42	mir-373	miR-373	gaagugcuucgauuuuggggugu
MIR374A	Xq13.2	mir-374a	miR-374a	uuauaauacaaccugauaagug
MIR374B	Xq13.2	mir-374b	miR-374b	auauaauacaaccugcuaagug

miRNA gene identifier	miRNA gene locus	miRNA precursor (stem-loop) identifier	mature miRNA identifier	mature miRNA sequence
MIR375	2q35	mir-375	miR-375	uuuguucguucggcucgcguga
MIR376A1	14q32.31	mir-376a-1	miR-376a	aucauagaggaaaauccacgu
MIR376A2	14q32.31	mir-376a-2	miR-376a	aucauagaggaaaauccacgu
MIR376B	14q32.31	mir-376b	miR-376b	aucauagaggaaaauccauguu
MIR376C	14q32.31	mir-376c	miR-376c	aacauagaggaaauuccacgu
MIR377	14q32.31	mir-377	miR-377	aucacacaaaggcaacuuuugu
MIR378	5q32	mir-378	miR-378	acuggacuuggagucagaagg
MIR379	14q32.31	mir-379	miR-379	ugguagacuauggaacguagg
MIR380	14q32.31	mir-380	miR-380	uauguaauaugguccacaucuu
MIR381	14q32.31	mir-381	miR-381	uauacaagggcaagcucucugu
MIR382	14q32.31	mir-382	miR-382	gaaguuguucgugguggauucg
MIR383	8p22	mir-383	miR-383	agaucagaaggugauuguggcu
MIR384	Xq21.1	mir-384	miR-384	auuccagaaauuguucaua
MIR409	14q32.31	mir-409	miR-409-5p	agguuacccgagcaacuuugcau
			miR-409-3p	gaaguugcucggugaaccccu
MIR410	14q32.31	mir-410	miR-410	aauauaacacagauggccugu
MIR411	14q32.31	mir-411	miR-411	uaguagaccguauagcguacg
MIR412	14q32.31	mir-412	miR-412	acuucaccugguccacuagccgu
MIR421	Xq13.2	mir-421	miR-421	aucaacagacauuaauugggcgc
MIR422A	15q22.2	mir-422a	miR-422a	acuggacuuagggucagaaggc
MIR423	17q11.2	mir-423	miR-423-5p	ugaggggcagagagcgagacuuu
			miR-423-3p	agcucggucugaggccccucagu
MIR424	Xq26.3	mir-424	miR-424	cagcagcaaucauguuuugaa
MIR425	3p21.31	mir-425	miR-425	aaugacacgaucacucccguuga
MIR429	1p36.33	mir-429	miR-429	uaauacugucugguaaaaccgu
MIR431	14q32.31	mir-431	miR-431	ugucuugcaggccgucaugca
MIR432	14q32.31	mir-432	miR-432	ucuuggaguaggucauugggugg
MIR433	14q32.31	mir-433	miR-433	aucaugaugggcuccucggugu
MIR448	Xq23	mir-448	miR-448	uugcauauguaggaugucccau
MIR449A	5q11.2	mir-449a	miR-449a	uggcaguguaauguuagcuggu
MIR449B	5q11.2	mir-449b	miR-449b	aggcaguguauuguuagcuggc
MIR449C	5q11.2	mir-449c	miR-449c	uaggcaguguauugcuagcggcugu
MIR450A1	Xq26.3	mir-450a-1	miR-450a	uuuugcgauguguuccuaauau
MIR450A2	Xq26.3	mir-450a-2	miR-450a	uuuugcgauguguuccuaauau
MIR450B	Xq26.3	mir-450b	miR-450b-5p	uuuugcaauauguuccugaaua
			miR-450b-3p	uugggaucauuuugcauccaua
MIR451	17q11.2	mir-451	miR-451	aaaccguuaccauuacugaguu
MIR452	Xq28	mir-452	miR-452	aacuguuugcagaggaaacuga
MIR453	14q32.31	mir-453	miR-453	agguuguccguggugaguucgca
MIR454	17q22	mir-454	miR-454	uagugcaauauugcuuauagggu
MIR455	9q32	mir-455	miR-455-5p	auaugccuuuggacuacaucg

miRNA gene identifier	miRNA gene locus	miRNA precursor (stem-loop) identifier	mature miRNA identifier	mature miRNA sequence
			miR-455-3p	gcaguccaugggcauauacac
MIR483	11p15.5	mir-483	miR-483-5p	aagacgggaggaaagaagggag
			miR-483-3p	ucacuccucuccucccgucuu
MIR484	16p13.11	mir-484	miR-484	ucaggcucaguccccucccgau
MIR485	14q32.31	mir-485	miR-485-5p	agaggcuggccgugaugaauuc
			miR-485-3p	gucauacacggcucuccucucu
MIR486	8p11.21	mir-486	miR-486-5p	uccuguacugagcugccccgag
			miR-486-3p	cggggcagcucaguacaggau
MIR487A	14q32.31	mir-487a	miR-487a	aaucauacagggacauccaguu
MIR487B	14q32.31	mir-487b	miR-487b	aaucguacagggucauccacuu
MIR488	1q25.2	mir-488	miR-488	uugaaaggcuauuucuuggc
MIR489	7q21.3	mir-489	miR-489	gugacaucacauauacggcagc
MIR490	7q33	mir-490	miR-490-5p	ccauggaucuccaggugggu
			miR-490-3p	caaccuggaggacuccaugcug
MIR491	9p21.3	mir-491	miR-491-5p	agugggaacccuuccaugagg
			miR-491-3p	cuuaugcaagauucccuucuac
MIR492	12q22	mir-492	miR-492	aggaccugcgggacaagauucuu
MIR493	14q32.31	mir-493	miR-493	ugaaggucuacugugugccagg
MIR494	14q32.31	mir-494	miR-494	ugaaacauacacgggaaaccuc
MIR495	14q32.31	mir-495	miR-495	aaacaaacauggugcacuucuu
MIR496	14q32.31	mir-496	miR-496	ugaguauuacauggccaaucuc
MIR497	17p13.1	mir-497	miR-497	cagcagcacacugugguuugu
MIR498	19q13.42	mir-498	miR-498	uuucaagccagggggcguuuuuc
MIR499	20q11.22	mir-499	miR-499-5p	uuaagacuugcagugauguuu
			miR-499-3p	aacaucacagcaagucugugcu
MIR500	Xp11.23	mir-500	miR-500	uaauccuugcuaccugggugaga
MIR501	Xp11.23	mir-501	miR-501-5p	aauccuuugucccugggugaga
			miR-501-3p	aaugcacccgggcaaggauucu
MIR502	Xp11.23	mir-502	miR-502-5p	auccuugcuaucugggugcua
			miR-502-3p	aaugcaccugggcaaggaauuca
MIR503	Xq26.3	mir-503	miR-503	uagcagcgggaacaguucugcag
MIR504	Xq26.3	mir-504	miR-504	agacccuggucugcacucuauc
MIR505	Xq27.1	mir-505	miR-505	cgucaacacuugcugguuuccu
MIR506	Xq27.3	mir-506	miR-506	uaaggcacccuucugaguaga
MIR507	Xq27.3	mir-507	miR-507	uuuugcaccuuuuggagugaa
MIR508	Xq27.3	mir-508	miR-508-5p	uacuccagagggcgucacucaug
			miR-508-3p	ugauuguagccuuuuggaguaga
MIR509-1	Xq27.3	mir-509-1	miR-509-5p	uacugcagacaguggcaauca
			miR-509-3p	ugauuggacgucuguggguag
MIR509-2	Xq27.3	mir-509-2	miR-509-5p	uacugcagacaguggcaauca
			miR-509-3p	ugauuggacgucuguggguag
MIR509-3	Xq27.3	mir-509-3	miR-509-3-5p	uacugcagacguggcaaucaug
			miR-509-3p	ugauuggacgucuguggguag
MIR510	Xq27.3	mir-510	miR-510	uacucaggagaguggcaaucac
MIR511-1	10p12.33	mir-511-1	miR-511	gugucuuuugcucugcaguca
MIR511-2	10p12.33	mir-511-2	miR-511	gugucuuuugcucugcaguca

miRNA gene identifier	miRNA gene locus	miRNA precursor (stem-loop) identifier	mature miRNA identifier	mature miRNA sequence
MIR512-1	19q13.42	mir-512-1	miR-512-5p	cacucagccuugagggcacuuuc
			miR-512-3p	aagugcugucauagcugagguc
MIR512-2	19q13.42	mir-512-2	miR-512-5p	cacucagccuugagggcacuuuc
			miR-512-3p	aagugcugucauagcugagguc
MIR513A1	Xq27.3	mir-513a-1	miR-513a-5p	uucacagggaggugucau
			miR-513a-3p	uaaauuucaccuuucugagaagg
MIR513A2	Xq27.3	mir-513a-2	miR-513a-5p	uucacagggaggugucau
			miR-513a-3p	uaaauuucaccuuucugagaagg
MIR513B	Xq27.3	mir-513b	miR-513b	uucacaaggaggugucauuuau
MIR513C	Xq27.3	mir-513c	miR-513c	uucucaaggaggugucguuuau
MIR514-1	Xq27.3	mir-514-1	miR-514	auugacacuucugugaguaga
MIR514-2	Xq27.3	mir-514-2	miR-514	auugacacuucugugaguaga
MIR514-3	Xq27.3	mir-514-3	miR-511	auugacacuucugugaguaga
MIR515-1	19q13.42	mir-515-1	miR-515-5p	uucuccaaaagaaagcacuuucug
			miR-515-3p	gagugccuucuuuuggagcguu
MIR515-2	19q13.42	mir-515-2	miR-515-5p	uucuccaaaagaaagcacuuucug
			miR-515-3p	gagugccuucuuuuggagcguu
MIR516A1	19q13.42	mir-516a-1	miR-516a-5p	uucucgaggaaagaagcacuuuc
			miR-516a-3p	ugcuuccuuucagagggu
MIR516A2	19q13.42	mir-516a-2	miR-516a-5p	uucucgaggaaagaagcacuuuc
			miR-516a-3p	ugcuuccuuucagagggu
MIR516B1	19q13.42	mir-516b-1	miR-516b	aucuggagguaagaagcacuuu
MIR516B2	19q13.42	mir-516b-2	miR-516b	aucuggagguaagaagcacuuu
MIR517A	19q13.42	mir-517a	miR-517a	aucgugcaucccuuuagagugu
MIR517B	19q13.42	mir-517b	miR-517b	ucgugcaucccuuuagaguguu
MIR517C	19q13.42	mir-517c	miR-517c	aucgugcauccuuuuagagugu
MIR518A1	19q13.42	mir-518a-1	miR-518a-5p	cugcaaagggaagcccuuuc
			miR-518a-3p	gaaagcgcuucccuuugcugga
MIR518A2	19q13.42	mir-518a-2	miR-518a-5p	cugcaaagggaagcccuuuc
			miR-518a-3p	gaaagcgcuucccuuugcugga
MIR518B	19q13.42	mir-518b	miR-518b	caaagcgcuccccuuuagaggu
MIR518C	19q13.42	mir-518c	miR-518c	caaagcgcuucucuuuagagugu
MIR518D	19q13.42	mir-518d	miR-518d-5p	cucuagagggaagcacuuucug
			miR-518d-3p	caaagcgcuucccuuuggagc
MIR518E	19q13.42	mir-518e	miR-518e	aaagcgcuucccuucagagug
MIR518F	19q13.42	mir-518f	miR-518f	gaaagcgcuucucuuuagagg
MIR519A1	19q13.42	mir-519a-1	miR-519a	aaagugcauccuuuuagagugu
MIR519A2	19q13.42	mir-519a-2	miR-519a	aaagugcauccuuuuagagugu
MIR519B	19q13.42	mir-519b	miR-519b-5p	cucuagagggaagcgcuuucug
			miR-519b-3p	aaagugcauccuuuuagagguu
MIR519C	19q13.42	mir-519c	miR-519c-5p	cucuagagggaagcgcuuucug
			miR-519c-3p	aaagugcaucuuuuagaggau
MIR519D	19q13.42	mir-519d	miR-519d	caaagugccucccuuuagagug
MIR519E	19q13.42	mir-519e	miR-519e	aagugccuccuuuuagaguguu
MIR520A	19q13.42	mir-520a	miR-520a-5p	cuccagagggaaguacuuucu
			miR-520a-3p	aaagugcuucccuuuggacugu
MIR520B	19q13.42	mir-520b	miR-520b	aaagugcuuccuuuuagaggg

miRNA gene identifier	miRNA gene locus	miRNA precursor (stem-loop) identifier	mature miRNA identifier	mature miRNA sequence
MIR520C	19q13.42	mir-520c	miR-520c-5p	cucuagagggaagcacuuucug
			miR-520c-3p	aaagugcuuccuuuuagagggu
MIR520D	19q13.42	mir-520d	miR-520d-5p	cuacaaagggaagcccuuuc
			miR-520d-3p	aaagugcuucucuuuggugggu
MIR520E	19q13.42	mir-520e	miR-520e	aaagugcuuccuuuuugaggg
MIR520F	19q13.42	mir-520f	miR-520f	aagugcuuccuuuuagaggguu
MIR520G	19q13.42	mir-520g	miR-520g	acaaagugcuucccuuuagagugu
MIR520H	19q13.42	mir-520h	miR-520h	acaaagugcuucccuuuagagu
MIR521-1	19q13.42	mir-521-1	miR-521	aacgcacuucccuuuagagugu
MIR521-2	19q13.42	mir-521-2	miR-521	aacgcacuucccuuuagagugu
MIR522	19q13.42	mir-522	miR-522	aaaauggucccuuuagagugu
MIR523	19q13.42	mir-523	miR-523	gaacgcgcuucccuauagagggu
MIR524	19q13.42	mir-524	miR-524-5p	cuacaaagggaagcacuuucuc
			miR-524-3p	gaaggcgcuuccccuuuggagu
MIR525	19q13.42	mir-525	miR-525-5p	cuccagagggaugcacuuucu
			miR-525-3p	gaaggcgcuucccuuuagagcg
MIR526A1	19q13.42	mir-526a-1	miR-526a	cucuagagggaagcacuuucug
MIR526A2	19q13.42	mir-526a-2	miR-526a	cucuagagggaagcacuuucug
MIR526B	19q13.42	mir-526b	miR-526b	cucuugagggaagcacuuucugu
MIR527	19q13.42	mir-527	miR-527	cugcaaagggaagcccuuuc
MIR532	Xp11.23	mir-532	miR-532-5p	caugccuugaguguaggaccgu
			miR-532-3p	ccucccacacccaaggcuugca
MIR539	14q32.31	mir-539	miR-539	ggagaaauuauccuuggugugu
MIR541	14q32.31	mir-541	miR-541	uggugggcacagaaucuggacu
MIR542	Xq26.3	mir-542	miR-542-5p	ucggggaucaucaugucacgaga
			miR-542-3p	ugugacagauugauaacugaaa
MIR543	14q32.31	mir-543	miR-543	aaacauucgcggugcacuucuu
MIR544	14q32.31	mir-544	miR-544	auucugcauuuuuagcaaguuc
MIR545	Xq13.2	mir-545	miR-545	ucagcaaacauuuauugugugc
MIR548A1	6p22.3	mir-548a-1	miR-548a-3p	caaaacuggcaauuacuuuugc
MIR548A2	6q23.3	mir-548a-2	miR-548a-3p	caaaacuggcaauuacuuuugc
MIR548A3	8q22.3	mir-548a-3	miR-548a-5p	aaaaguaauugcgaguuuuacc
			miR-548a-3p	caaaacuggcaauuacuuuugc
MIR548B	6q22.31	mir-548b	miR-548b-5p	aaaaguaauugugguuuuggcc
			miR-548b-3p	caagaaccucaguugcuuuugu
MIR548C	12q14.2	mir-548c	miR-548c-5p	aaaaguaauugcgguuuuugcc
			miR-548c-3p	caaaaaucucaauuacuuuugc
MIR548D1	8q24.13	mir-548d-1	miR-548d-5p	aaaaguaauugugguuuuugcc
			miR-548d-3p	caaaaaccacaguuucuuuugc
MIR548D2	17q24.2	mir-548d-2	miR-548d-5p	aaaaguaauugugguuuuugcc
			miR-548d-3p	caaaaaccacaguuucuuuugc
MIR548E	10q25.2	mir-548e	miR-548e	aaaaacugagacuacuuuugca
MIR548F1	10q21.1	mir-548f-1	miR-548f	aaaaacuguaauuacuuuu

miRNA gene identifier	miRNA gene locus	miRNA precursor (stem-loop) identifier	mature miRNA identifier	mature miRNA sequence
MIR548F2	2q34	mir-548f-2	miR-548f	aaaaacuguaauuacuuuu
MIR548F3	5q22.1	mir-548f-3	miR-548f	aaaaacuguaauuacuuuu
MIR548F4	7q35	mir-548f-4	miR-548f	aaaaacuguaauuacuuuu
MIR548F5	Xp21.1	mir-548f-5	miR-548f	aaaaacuguaauuacuuuu
MIR548G	4q31.22	mir-548g	miR-548g	aaaacuguaauuacuuuuguac
MIR548H1	14q23.2	mir-548h-1	miR-548h	aaaaguaaucgcgguuuuuguc
MIR548H2	16p13.13	mir-548h-2	miR-548h	aaaaguaaucgcgguuuuuguc
MIR548H3	17p12	mir-548h-3	miR-548h	aaaaguaaucgcgguuuuuguc
MIR548H4	8p21.2	mir-548h-4	miR-548h	aaaaguaaucgcgguuuuuguc
MIR548I1	3q21.2	mir-548i-1	miR-548i	aaaaguaauugcggauuuugcc
MIR548I2	4p16.1	mir-548i-2	miR-548i	aaaaguaauugcggauuuugcc
MIR548I3	8p23.1	mir-548i-3	miR-548i	aaaaguaauugcggauuuugcc
MIR548I4	Xq21.1	mir-548i-4	miR-548i	aaaaguaauugcggauuuugcc
MIR548J	22q12.1	mir-548j	miR-548j	aaaaguaaaugcggucuuuggu
MIR548K	11q13.3	mir-548k	miR-548k	aaaaguacuugcggauuuugcu
MIR548L	11q21	mir-548l	miR-548l	aaaaguauuugcgguuuuuguc
MIR548M	Xq21.33	mir-548m	miR-548m	caaaggUauuugugguuuuug
MIR548N	7p14.3	mir-548n	miR-548n	caaaaguaauugUggauuuugu
MIR548O	7q22.1	mir-548o	miR-548o	ccaaaacugcaguuacuuuugc
MIR548P	5q21.1	mir-548p	miR-548p	uagcaaaacugcaguuacuuu
MIR548Q	10p13	mir-548q	miR-548q	gcuggugcaaaaguaauggcgg
MIR549	15q25.1	mir-549	miR-549	ugacaacuauggaugagcucu
MIR550-1	7p14.3	mir-550-1	miR-550	agugccugagggaguaagagccc
MIR550-2	7p14.3	mir-550-2	miR-550	agugccugagggaguaagagccc
MIR551A	1p36.32	mir-551a	miR-551a	gcgacccacucugguuucca
MIR551B	3q26.2	mir-551b	miR-551b	gcgacccauacugguuucag
MIR552	1p34.3	mir-552	miR-552	aacaggugacugguuagacaa
MIR553	1p21.2	mir-553	miR-553	aaaacggugagauuuuguuuu
MIR554	1q21.3	mir-554	miR-554	gcuaguccugacucagccagu
MIR555	1q22	mir-555	miR-555	agggUaagcugaaccucugau
MIR556	1q23.3	mir-556	miR-556-5p	gaugagcucauuguaauaugag
			miR-556-3p	auauuaccauuagcucaucuuu
MIR557	1q24.2	mir-557	miR-557	guuugcacggugggccuugucu
MIR558	2p22.3	mir-558	miR-558	ugagcugcuguaccaaaau
MIR559	2p21	mir-559	miR-559	uaaaguaaauaugcaccaaaa
MIR561	2q32.1	mir-561	miR-561	caaaguuuaagauccuugaagu
MIR562	2q37.1	mir-562	miR-562	aaaguagcuguaccauuugc
MIR563	3p25.1	mir-563	miR-563	agguugacauacguuuccc
MIR564	3p21.31	mir-564	miR-564	aggcacggugucagcaggc
MIR566	3p21.31	mir-566	miR-566	gggcgccugugaucccaac
MIR567	3q13.2	mir-567	miR-567	aguauguucuuccaggacagaac
MIR568	3q13.31	mir-568	miR-568	auguauaaauguauacacac
MIR569	3q26.2	mir-569	miR-569	aguuaaugaauccuggaaagu
MIR570	3q29	mir-570	miR-570	cgaaaacagcaauuaccuuugc
MIR571	4p16.3	mir-571	miR-571	ugaguuggccaucugagugag

Human MiRNAs: Genes, Names, Loci, Sequences, Clusters

miRNA gene identifier	miRNA gene locus	miRNA precursor (stem-loop) identifier	mature miRNA identifier	mature miRNA sequence
MIR572	4p15.33	mir-572	miR-572	guccgcucggcgguggccca
MIR573	4p15.2	mir-573	miR-573	cugaagugauguguaacugaucag
MIR574	4p14	mir-574	miR-574-5p	ugagugugugugugugagugugu
			miR-574-3p	cacgcucaugcacacacccaca
MIR575	4q21.22	mir-575	miR-575	gagccaguuggacaggagc
MIR576	4q25	mir-576	miR-576-5p	auucuaauuucuccacgucuuu
			miR-576-3p	aagauguggaaaaauuggaauc
MIR577	4q26	mir-577	miR-577	uagauaaaauauugguaccug
MIR578	4q32.3	mir-578	miR-578	cuucuugugcucuaggauugu
MIR579	5p13.3	mir-579	miR-579	uucauuugguauaaaccgcgauu
MIR580	5p13.2	mir-580	miR-580	uugagaaugaugaaucauuagg
MIR581	5q11.2	mir-581	miR-581	ucuuguguucucuagaucagu
MIR582	5q12.1	mir-582	miR-582-5p	uuacaguuguucaaccaguuacu
			miR-582-3p	uaacugguugaacaacugaacc
MIR583	5q15	mir-583	miR-583	caaagaggaaggucccauuac
MIR584	5q32	mir-584	miR-584	uuauggnuugccugggacugag
MIR585	5q35.1	mir-585	miR-585	ugggcguaucuguaugcua
MIR586	6p21.1	mir-586	miR-586	uaugcauuguauuuuuaggucc
MIR587	6q21	mir-587	miR-587	uuuccauaggugaugagucac
MIR588	6q22.32	mir-588	miR-588	uuggccacaauggguuagaac
MIR589	7p22.1	mir-589	miR-589	ugagaaccacgucugcucugag
MIR590	7q11.23	mir-590	miR-590-5p	gagcuuauucauaaaagugcag
			miR-590-3p	uaauuuuauguauaagcuagu
MIR591	7q21.3	mir-591	miR-591	agaccaugggunucucauugu
MIR592	7q31.33	mir-592	miR-592	uugugucaauaugcgaugaugu
MIR593	7q32.1	mir-593	miR-593	ugucucugcugggguuucu
MIR595	7q36.3	mir-595	miR-595	gaagugugccguggugugucu
MIR596	8p23.3	mir-596	miR-596	aagccugcccggcuccucggg
MIR597	8p23.1	mir-597	miR-597	ugugucacucgaugaccacugu
MIR598	8p23.1	mir-598	miR-598	uacgucaucguugucaucguca
MIR599	8q22.2	mir-599	miR-599	guugugucaguuuaucaaac
MIR600	9q33.3	mir-600	miR-600	acuuacagacaagagccuugcuc
MIR601	9q33.3	mir-601	miR-601	uggucuaggauuguuggaggag
MIR602	9q34.3	mir-602	miR-602	gacacgggcgacagcugcggccc
MIR603	10p12.2	mir-603	miR-603	cacacacugcaauuacuuuugc
MIR604	10p11.23	mir-604	miR-604	aggcugcggaauucaggac
MIR605	10q21.1	mir-605	miR-605	uaaaucccauggugccuucuccu
MIR606	10q22.2	mir-606	miR-606	aaacuacugaaaaucaaagau
MIR607	10q24.1	mir-607	miR-607	guucaaauccagaucuauaac
MIR608	10q24.31	mir-608	miR-608	aggggugguguugggacagcuccgu
MIR609	10q25.1	mir-609	miR-609	aggguguuucucucaucucu
MIR610	11p14.1	mir-610	miR-610	ugagcuaaaugugugcuggga
MIR611	11q12.2	mir-611	miR-611	gcgaggaccccucggggucugac
MIR612	11q13.1	mir-612	miR-612	gcugggcagggcuucugagcuccuu

miRNA gene identifier	miRNA gene locus	miRNA precursor (stem-loop) identifier	mature miRNA identifier	mature miRNA sequence
MIR613	12p13.1	mir-613	miR-613	aggaauguuccuucuuugcc
MIR614	12p13.1	mir-614	miR-614	gaacgccuguucuugccaggugg
MIR615	12q13.13	mir-615	miR-615-5p	ggggguccccggugcucggauc
			miR-615-3p	uccgagccugggucucccucuu
MIR616	12q13.3	mir-616	miR-616	agucauuggaggguuugagcag
MIR617	12q21.31	mir-617	miR-617	agacuucccauuugaagguggc
MIR618	12q21.31	mir-618	miR-618	aaacucuacuuguccuucugagu
MIR619	12q24.11	mir-619	miR-619	gaccuggacauguuugugcccagu
MIR620	12q24.21	mir-620	miR-620	auggagauagauauagaaau
MIR621	13q14.11	mir-621	miR-621	ggcuagcaacagcgcuuaccu
MIR622	13q31.2	mir-622	miR-622	acagucugcugagguuggagc
MIR623	13q32.3	mir-623	miR-623	aucccuugcaggggcuguugggu
MIR624	14q12	mir-624	miR-624	cacaagguaauugguauuaccu
MIR625	14q23.3	mir-625	miR-625	aggggggaaaguucuauagucc
MIR626	15q15.1	mir-626	miR-626	agcugucugaaaaugucuu
MIR627	15q15.1	mir-627	miR-627	gugagucucuaagaaaagagga
MIR628	15q21.3	mir-628	miR-628-5p	augcugacauauuuacuagagg
			miR-628-3p	ucuaguaagaguggcagucga
MIR629	15q23	mir-629	miR-629	ugggUuuacguugggagaacu
MIR630	15q24.1	mir-630	miR-630	aguauucuguaccagggaaggu
MIR631	15q24.2	mir-631	miR-631	agaccuggcccagaccucagc
MIR632	17q11.2	mir-632	miR-632	gugucugcuuccugug gga
MIR633	17q23.2	mir-633	miR-633	cuaauaguaucuaccacaauaaa
MIR634	17q24.2	mir-634	miR-634	aaccagcaccccaacuuuggac
MIR635	17q24.2	mir-635	miR-635	acuugggcacugaaacaaugcc
MIR636	17q25.1	mir-636	miR-636	ugugcuugcucgucccgcccgca
MIR637	19p13.3	mir-637	miR-637	acuggggcuuucgggcucugcgu
MIR638	19p13.2	mir-638	miR-638	agggaucgcgggcgguggcggccu
MIR639	19p13.12	mir-639	miR-639	aucgcugcgguugcgagcgcugu
MIR640	19p13.11	mir-640	miR-640	augauccaggaaccugccucu
MIR641	19q13.2	mir-641	miR-641	aaagacauaggauagagucaccuc
MIR642	19q13.32	mir-642	miR-642	gucccucuccaaaugucuug
MIR643	19q13.41	mir-643	miR-643	acuuguaugcuagcucagguag
MIR644	20q11.22	mir-644	miR-644	aguuggcuuucuuagagc
MIR645	20q13.13	mir-645	miR-645	ucuaggcugguacugcuga
MIR646	20q13.33	mir-646	miR-646	aagcagcugccucugaggc
MIR647	20q13.33	mir-647	miR-647	gugg cugcacucacuuccuuc
MIR648	22q11.21	mir-648	miR-648	aagugugcagggcacuggu
MIR649	22q11.21	mir-649	miR-649	aaaccuguguuguucaagaguc
MIR650	22q11.22	mir-650	miR-650	aggaggcagcgcucucaggac
MIR651	Xp22.31	mir-651	miR-651	uuuaggauaagcuugacuuuug
MIR652	Xq23	mir-652	miR-652	aauggcgccacuaggguugug
MIR653	7q21.3	mir-653	miR-653	guguugaaacaaucucuacug
MIR654	14q32.31	mir-654	miR-654-5p	uggugggccgcagaacaugugc
			miR-654-3p	uaugucugcugaccaucaccuu
MIR655	14q32.31	mir-655	miR-655	auaauacagguuaaccucuuu

miRNA gene identifier	miRNA gene locus	miRNA precursor (stem-loop) identifier	mature miRNA identifier	mature miRNA sequence
MIR656	14q32.31	mir-656	miR-656	aauauuauacagucaaccucu
MIR657	17q25.3	mir-657	miR-657	ggcagguucucacccucucuagg
MIR658	22q13.1	mir-658	miR-658	ggcggagggaaguagguccguugg u
MIR659	22q13.1	mir-659	miR-659	cuugguucagggagggucccca
MIR660	Xp11.23	mir-660	miR-660	uacccauugcauaucggaguug
MIR661	8q24.3	mir-661	miR-661	ugccugggucucuggccugcgcgu
MIR662	16p13.3	mir-662	miR-662	ucccacguuguggcccagcag
MIR663	20ph (the heterochromatic region on the p arm)	mir-663	miR-663	aggcggggcgccgcgggaccgc
MIR663B	2q21.2	mir-663b	miR-663b	gguggcccggccgugccugagg
MIR664	1q41	mir-664	miR-664	uauucauuuaucccagccuaca
MIR665	14q32.31	mir-665	miR-665	accaggaggcugaggccccu
MIR668	14q32.31	mir-668	miR-668	ugucacucggcucggcccacuac
MIR670	11p11.2	mir-670	miR-670	gucccugaguguauguggug
MIR671	7q36.1	mir-671	miR-671-5p	aggaagcccuggaggggcuggag
			miR-671-3p	uccgguucucagggcuccacc
MIR675	11p15.5	mir-675	miR-675	uggugcggagagggcccacagug
MIR708	11q14.1	mir-708	miR-708	aaggagcuuacaaucuagcuggg
MIR711	3p21.31	mir-711	miR-711	gggacccagggagagacguaag
MIR718	Xq28	mir-718	miR-718	cuuccgccccgccgggcgucg
MIR720	3q26.1	mir-720	miR-720	ucucgcuggggccucca
MIR744	17p12	mir-744	miR-744	ugcggggcuagggcuaacagca
MIR758	14q32.31	mir-758	miR-758	uuugugaccugguccacuaacc
MIR759	13q14.3	mir-759	miR-759	gcagagugcaaacaauuuugac
MIR760	1p22.1	mir-760	miR-760	cggcucugggucugugggga
MIR761	1p32.3	mir-761	miR-761	gcagcagggugaaacugacaca
MIR762	16p11.2	mir-762	miR-762	ggggcuggggccggggccgagc
MIR764	Xq23	mir-764	miR-764	gcaggugcucacuuguccuccu
MIR765	1q23.1	mir-765	miR-765	uggaggagaaggaaggugaug
MIR766	Xq24	mir-766	miR-766	acuccagcccacagccucagc
MIR767	Xq28	mir-767	miR-767-5p	ugcaccaugguugucugagcaug
			miR-767-3p	ucugcucauaccccaugguuucu

miRNA gene identifier	miRNA gene locus	miRNA precursor (stem-loop) identifier	mature miRNA identifier	mature miRNA sequence
MIR769	19q13.32	mir-769	miR-769-5p	ugagaccucugggúucugagcu
			miR-769-3p	cugggaucuccggggggucuugguu
MIR770	14q32.31	mir-770	miR-770-5p	uccaguaccacgugucagggcca
MIR802	21q22.12	mir-802	miR-802	caguaacaaagauucauccuugu
MIR873	9p21.1	mir-873	miR-873	gcaggaacuugugagucuccu
MIR874	5q31.2	mir-874	miR-874	cugcccuggcccgagggaccga
MIR875	8q22.2	mir-875	miR-875-5p	uauaccucaguuuuaucaggug
			miR-875-3p	ccuggaaacacugagguugug
MIR876	9p21.1	mir-876	miR-876-5p	uggauuucuuugugaaucacca
			miR-876-3p	uggugguuuacaaaguaauuca
MIR877	6p21.33	mir-877	miR-877	guagaggagauggcgcaggg
MIR885	3p25.3	mir-885	miR-885-5p	uccauuacacuacccugccucu
			miR-885-3p	aggcagcggggguguagugg aua
MIR886	5q31.1	mir-886	miR-886-5p	cgggucggaguuagcucaagcgg
			miR-886-3p	cgcgggugcuuacugacccuu
MIR887	5p15.1	mir-887	miR-887	gugaacgggcgccaucccgagg
MIR888	Xq27.3	mir-888	miR-888	uacucaaaaagcugucaguca
MIR889	14q32.31	mir-889	miR-889	uuaauaucggacaaccauugu
MIR890	Xq27.3	mir-890	miR-890	uacuuggaaaggcaucaguug
MIR891A	Xq27.3	mir-891a	miR-891a	ugcaacgaaccugagccacuga
MIR891B	Xq27.3	mir-891b	miR-891b	ugcaacuuaccugagucauuga
MIR892A	Xq27.3	mir-892a	miR-892a	cacuguguccuuucugcguag
MIR892B	Xq27.3	mir-892b	miR-892b	cacuggcuccuuucugguaga
MIR920	12p12.1	mir-920	miR-920	ggggagcuguggaagcagua
MIR921	1q24.1	mir-921	miR-921	cuagugagggacagaaccaggauuc
MIR922	3q29	mir-922	miR-922	gcagcagagaauaggacuacguc
MIR924	18q12.3	mir-924	miR-924	agagucuugugaugucuugc
MIR933	2q31.1	mir-933	miR-933	ugugcgcagggagaccucuccc
MIR934	Xq26.3	mir-934	miR-934	ugucuacuacuggagacacugg
MIR935	19q13.42	mir-935	miR-935	ccaguuaccgcuuccgcuaccgc
MIR936	10q25.1	mir-936	miR-936	acaguagaggggaggaaucgcag
MIR937	8q24.3	mir-937	miR-937	auccgcgcucugacucucugcc
MIR938	10p11.23	mir-938	miR-938	ugcccuuaaaggugaaccagu
MIR939	8q24.3	mir-939	miR-939	uggggagcugaggcucuggggug
MIR940	16p13.3	mir-940	miR-940	aaggcagggcccccgcuccccc
MIR941-1	20q13.33	mir-941-1	miR-941	cacccggcugugugcacaugugc
MIR941-2	20q13.33	mir-941-2	miR-941	cacccggcugugugcacaugugc
MIR941-3	20q13.33	mir-941-3	miR-941	cacccggcugugugcacaugugc
MIR941-4	20q13.33	mir-941-4	miR-941	cacccggcugugugcacaugugc
MIR942	1p13.1	mir-942	miR-942	ucuucucuguuuuggccaugug
MIR943	4p16.3	mir-943	miR-943	cugacuguugccguccuccag

miRNA gene identifier	miRNA gene locus	miRNA precursor (stem-loop) identifier	mature miRNA identifier	mature miRNA sequence
MIR944	3q28	mir-944	miR-944	aaauuauuguacaucggaugag
MIR1178	12q24.23	mir-1178	miR-1178	uugcucacuguucuucccuag
MIR1179	15q26.1	mir-1179	miR-1179	aagcauucuuucauugguugg
MIR1180	17p11.2	mir-1180	miR-1180	uuuccggcucgcguggguguu
MIR1181	19p13.2	mir-1181	miR-1181	ccgucgccgccacccgagccg
MIR1182	1q42.2	mir-1182	miR-1182	gagggucuugggagggaugugac
MIR1183	7p15.3	mir-1183	miR-1183	cacuguaggugauggugagagugggca
MIR1184	Xq28	mir-1184	miR-1184	ccugcagcgacuugauggcuucc
MIR1185-1	14q32.31	mir-1185-1	miR-1185	agaggauacccuuuguauguu
MIR1185-2	14q32.31	mir-1185-2	miR-1185	agaggauacccuuuguauguu
MIR1197	14q32.31	mir-1197	miR-1197	uaggacacauggucuacuucu
MIR1200	7p14.2	mir-1200	miR-1200	cuccugagccauucugagccuc
MIR1201	14q11.2	mir-1201	miR-1201	agccugauuaaacacaugcucuga
MIR1202	6q25.3	mir-1202	miR-1202	gugccagcugcagugggggag
MIR1203	17q21.32	mir-1203	miR-1203	cccggagccaggaugcagcuc
MIR1204	8q24.21	mir-1204	miR-1204	ucguggccuggucuccauuau
MIR1205	8q24.21	mir-1205	miR-1205	ucugcagggpuugcuuugag
MIR1206	8q24.21	mir-1206	miR-1206	uguucauguagauguuuaagc
MIR1207	8q24.21	mir-1207	miR-1207-5p	uggcagggaggcugggaggg
			miR-1207-3p	ucagcuggcccucauuuc
MIR1208	8q24.21	mir-1208	miR-1208	ucacuguucagacaggcgga
MIR1224	3q27.1	mir-1224	miR-1224-5p	gugaggacucgggaggugg
			miR-1224-3p	ccccaccuccucucuccucag
MIR1225	16p13.3	mir-1225	miR-1225-5p	guggguacggcccagugggggg
			miR-1225-3p	ugagcccugugccgcccccag
MIR1226	3p21.31	mir-1226	miR-1226	ucaccagcccuguguucccuag
MIR1227	19p13.3	mir-1227	miR-1227	cgugccacccuuuuucccag
MIR1228	12q13.3	mir-1228	miR-1228	ucacaccugccucgcccccc
MIR1229	5q35.3	mir-1229	miR-1229	cucucaccacugcccuccacag
MIR1231	1q32.1	mir-1231	miR-1231	gugucugggcggacagcugc
MIR1233	15q14	mir-1233	miR-1233	ugagcccuguccucccgcag
MIR1234	8q24.3	mir-1234	miR-1234	ucggccugaccacccaccccac
MIR1236	6p21.33	mir-1236	miR-1236	ccucuuccccuugucucuccag
MIR1237	11q13.1	mir-1237	miR-1237	uccuucugcuccguccccag
MIR1238	19p13.2	mir-1238	miR-1238	cuuccucgucugcccc
MIR1243	4q25	mir-1243	miR-1243	aacuggaucaauuauaggagug
MIR1244	2q37.1, 5q23.1	mir-1244	miR-1244	aaguaguugguuuguaugagauggguu

miRNA gene identifier	miRNA gene locus	miRNA precursor (stem-loop) identifier	mature miRNA identifier	mature miRNA sequence
	12p13.31			
	12p13.2			
MIR1245	2q32.2	mir-1245	miR-1245	aagugaucuaaaggccuacau
MIR1246	2q31.1	mir-1246	miR-1246	aauggauuuuuggagcagg
MIR1247	14q32.31	mir-1247	miR-1247	acccgucccguucgucccga
MIR1248	3q27.3	mir-1248	miR-1248	accuucuuguauaagcacugugcuaaa
MIR1249	22q13.31	mir-1249	miR-1249	acgcccuucccccccuucuuca
MIR1250	17q25.3	mir-1250	miR-1250	acggugcuggaugugggccuuu
MIR1251	12q23.1	mir-1251	miR-1251	acucuagcugccaaaggcgcu
MIR1252	12q21.2	mir-1252	miR-1252	agaaggaaauugaauucauuua
MIR1253	17p13.3	mir-1253	miR-1253	agagaagaagaucagccugca
MIR1254	10q21.3	mir-1254	miR-1254	agccuggaagcuggagccugcagu
MIR1255A	4q14	mir-1255a	miR-1255a	aggaugagcaaagaaaguagauu
MIR1255B1	4p14	mir-1255b-1	miR-1255b	cggaugagcaaagaaaguggguu
MIR1255B2	1q24.2	mir-1255b-2	miR-1255b	cggaugagcaaagaaaguggguu
MIR1256	1p36.12	mir-1256	miR-1256	aggcauugacuucucacuagcu
MIR1257	20q13.33	mir-1257	miR-1257	agugaaugauggguucugacc
MIR1258	2q31.3	mir-1258	miR-1258	aguuaggauuaggucguggaa
MIR1259	20q13.13	mir-1259	miR-1259	auauaugaugacuuagcuuuu
MIR1260	14q24.3	mir-1260	miR-1260	aucccaccucugccacca
MIR1261	11q14.3	mir-1261	miR-1261	auggauaaggcuuuggcuu
MIR1262	1p31.3	mir-1262	miR-1262	augggugaauuuguagaaggau
MIR1263	3q26.1	mir-1263	miR-1263	augguacccuggcauacugagu
MIR1264	Xq23	mir-1264	miR-1264	caagucuuauuugagcaccuguu
MIR1265	10p13	mir-1265	miR-1265	caggaugugggucaaguguuguu
MIR1266	15q21.2	mir-1266	miR-1266	ccucagggcuguagaacagggcu
MIR1267	13q33.3	mir-1267	miR-1267	ccuguugaaguguaaucccca
MIR1268	15q11.2	mir-1268	miR-1268	cgggcguggugguggggg
MIR1269	4q13.2	mir-1269	miR-1269	cuggacugagccgugcuacugg
MIR1270	19p12	mir-1270	miR-1270	cuggagauauggaagagcugugu
MIR1271	5q35.2	mir-1271	miR-1271	cuuggcaccuagcaagcacuca
MIR1272	15q22.31	mir-1272	miR-1272	gaugaugauggcagcaaauucugaaa
MIR1273	8q22.2	mir-1273	miR-1273	gggcgacaaagcaagacucuuucuu
MIR1274A	5p13.1	mir-1274a	miR-1274a	gucccguucaggcgcca
MIR1274B	19q13.43	mir-1274b	miR-1274b	ucccuguucgggcgcca
MIR1275	6p21.31	mir-1275	miR-1275	guggggagaggcuguc
MIR1276	15q25.3	mir-1276	miR-1276	uaaagagcccuguggagaca
MIR1277	Xq24	mir-1277	miR-1277	uacgagauauauauguauuuu
MIR1278	1q31.2	mir-1278	miR-1278	uaguacugugcauaucaucuau
MIR1279	12q15	mir-1279	miR-1279	ucauauugcuucuuucu
MIR1280	3q21.3	mir-1280	miR-1280	ucccaccgcugccaccc
MIR1281	22q13.2	mir-1281	miR-1281	ucgccuccuccucuccc
MIR1282	15q15.3	mir-1282	miR-1282	ucguuugccuuuuucugcuu

miRNA gene identifier	miRNA gene locus	miRNA precursor (stem-loop) identifier	mature miRNA identifier	mature miRNA sequence
MIR1283-1	19q13.42	mir-1283-1	miR-1283	ucuacaaaggaaagcgcuuucu
MIR1283-2	19q13.42	mir-1283-2	miR-1283	ucuacaaaggaaagcgcuuucu
MIR1284	3p13	mir-1284	miR-1284	ucuauacagacccuggcuuuuc
MIR1285-1	7q21.2	mir-1285-1	miR-1285	ucugggcaacaaagugagaccu
MIR1285-2	2p13.3	mir-1285-2	miR-1285	ucugggcaacaaagugagaccu
MIR1286	22q11.21	mir-1286	miR-1286	ugcaggaccaagaugagcccu
MIR1287	10q24.2	mir-1287	miR-1287	ugcuggaucagugguucgaguc
MIR1288	17p11.2	mir-1288	miR-1288	uggacugcccugaucuggaga
MIR1289-1	20q11.22	mir-1289-1	miR-1289	uggaguccaggaaucugcauuuu
MIR1289-2	5q31.1	mir-1289-2	miR-1289	uggaguccaggaaucugcauuuu
MIR1290	1p36.13	mir-1290	miR-1290	uggauuuuuggaucaggga
MIR1291	12q13.11	mir-1291	miR-1291	uggcccugacugaagaccagcagu
MIR1292	20p13	mir-1292	miR-1292	ugggaacggguuccggcagacgcug
MIR1293	12q13.12	mir-1293	miR-1293	ugggguggucuggagauuugugc
MIR1294	5q33.2	mir-1294	miR-1294	ugugagguuggcauuguugucu
MIR1295	1q24.3	mir-1295	miR-1295	uuaggccgcagaucuggguga
MIR1296	10q21.3	mir-1296	miR-1296	uuagggcccuggcuccaucucc
MIR1297	13q14.3	mir-1297	miR-1297	uucaaguaauucaggug
MIR1298	Xq23	mir-1298	miR-1298	uucauucggcuguccagaugua
MIR1299	9q21.11	mir-1299	miR-1299	uucuggaauucugugugaggga
MIR1300	15q21.2	mir-1300	miR-1300	uugagaaggaggcugcug
MIR1301	2p23.3	mir-1301	miR-1301	uugcagcugccugggagugacuuc
MIR1302-1	12q24.13	mir-1302-1	miR-1302	uugggacauacuuaugcuaaa
MIR1302-2	1p36.33 9p24.3 15q26.3 19p13.3	mir-1302-2	miR-1302	uugggacauacuuaugcuaaa
MIR1302-3	2q13	mir-1302-3	miR-1302	uugggacauacuuaugcuaaa
MIR1302-4	2q33.3	mir-1302-4	miR-1302	uugggacauacuuaugcuaaa
MIR1302-5	20q13.13	mir-1302-5	miR-1302	uugggacauacuuaugcuaaa
MIR1302-6	7p21.1	mir-1302-6	miR-1302	uugggacauacuuaugcuaaa
MIR1302-7	8q24.3	mir-1302-7	miR-1302	uugggacauacuuaugcuaaa
MIR1302-8	9q22.33	mir-1302-8	miR-1302	uugggacauacuuaugcuaaa
MIR1303	5q33.2	mir-1303	miR-1303	uuuagagacggggucuugcucu
MIR1304	11q21	mir-1304	miR-1304	uuugaggcuacagugagaugug
MIR1305	4q34.3	mir-1305	miR-1305	uuuucaacucuaaugggagaga
MIR1306	22q11.21	mir-1306	miR-1306	acucggcguggcgucggucgug
MIR1307	10q24.33	mir-1307	miR-1307	acucggcguggcgucggucgug
MIR1308	Xp22.11	mir-1308	miR-1308	gcaugggugguucagugg
MIR1321	Xq21.2	mir-1321	miR-1321	cagggaggugaaugugau
MIR1322	8p23.1	mir-1322	miR-1322	gaugaugcugcugaugcug
MIR1323	19q13.42	mir-1323	miR-1323	ucaaaacugagggcauuuucu
MIR1324	3p12.3	mir-1324	miR-1324	ccagacagaauucuaugcacuuuc
MIR1468	Xq11.2	mir-1468	miR-1468	cuccguuugccuguuucgcug

miRNA gene identifier	miRNA gene locus	miRNA precursor (stem-loop) identifier	mature miRNA identifier	mature miRNA sequence
MIR1469	15q26.2	mir-1469	miR-1469	cucggcgcggggcgcgggcucc
MIR1470	19p13.12	mir-1470	miR-1470	gcccuccgcccgugcaccccg
MIR1471	2q37.1	mir-1471	miR-1471	gcccgcguguggagccaggugu
MIR1537	1q42.3	mir-1537	miR-1537	aaaaccgucuaguuacaguugu
MIR1538	16q22.1	mir-1538	miR-1538	cggcccgggcugcugcuguuccu
MIR1539	18q21.1	mir-1539	miR-1539	uccugcgcgucccagaugccc
MIR1825	20q11.21	mir-1825	miR-1825	uccagugcccuccucucc
MIR1826	16pter-qter	mir-1826	miR-1826	auugaucaucgacacuucgaacgcaau
MIR1827	12q23.1	mir-1827	miR-1827	ugaggcaguagauugaau
MIR1908	11q12.2	mir-1908	miR-1908	cggcgggacggcgauugguc
MIR1909	19p13.3	mir-1909	miR-1909	cgcaggggccgggugcucaccg
MIR1910	16q24.1	mir-1910	miR-1910	ccaguccugugccugccgccu
MIR1911	Xq23	mir-1911	miR-1911	ugaguaccgccaugucuguuggg
MIR1912	Xq23	mir-1912	miR-1912	uacccagagcaugcagugugaa
MIR1913	6q27	mir-1913	miR-1913	ucugcccccuccgcugcugcca
MIR1914	20q13.33	mir-1914	miR-1914	cccugugcccggcccacuucug
MIR1915	10p12.31	mir-1915	miR-1915	ccccagggcgacgcggcggg
MIR1972	16p13.11 16q22.1	mir-1972	miR-1972	ucaggccaggcacaguggcuca
MIR1973	4q26	mir-1973	miR-1973	accgugcaaagguagcaua
MIR1974	5q15 Chromosome MT	mir-1974	miR-1974	ugguuguaguccgugcgagaaua
MIR1975	7q36.1	mir-1975	miR-1975	cccccacaaccgcgcuugacuagcu
MIR1976	1p36.11	mir-1976	miR-1976	ccuccugcccuccuugcugu
MIR1977	1p36.33 Chromosome MT	mir-1977	miR-1977	gauuagggugcuuagcuguuaa
MIR1978	2q23.1 Chromosome MT	mir-1978	miR-1978	gguuugguccuagccuuucua
MIR1979	4q32.3	mir-1979	miR-1979	cucccacugcuucacuugacua
MIR2052	8q21.11	mir-2052	miR-2052	uguuuugauaacaguaaugu
MIR2053	8q23.3	mir-2053	miR-2053	guguuaauuaaaccucuauuuac
MIR2054	4q28.1	mir-2054	miR-2054	cuguaauauaaauuuaauuuauu
MIR2110	10q25.3	mir-2110	miR-2110	uuggggaaacggccgcugagug
MIR2113	6q16.1	mir-2113	miR-2113	auuugugcuuggcucugucac
MIR2114	Xq28	mir-2114	miR-2114	uagucccuuccuugaagcgguc
MIR2115	3p21.31	mir-2115	miR-2115	agcuuccaugacuccugaugga

miRNA gene identifier	miRNA gene locus	miRNA precursor (stem-loop) identifier	mature miRNA identifier	mature miRNA sequence
MIR2116	15q22.2	mir-2116	miR-2116	gguucuuagcauaggaggucu
MIR2117	17q21.31	mir-2117	miR-2117	uguucucuuugccaaggacag
MIR2276	13q12.12	mir-2276	miR-2276	ucugcaagugucagaggcgagg
MIR2277	5q15	mir-2277	miR-2277	ugacagcgcccugccuggcuc
MIR2278	9q22.32	mir-2278	miR-2278	gagagcaguguguguugccugg

Table 2. List of miRNA clusters

Locus	Cluster members
1p34.2	MIR30C1, MIR30E
1p36.33	MIR200A, MIR200B, MIR429
1q24.3	MIR199A2, MIR214
1q31.3	MIR181A1, MIR181B1
1q32.2	MIR29B-2, MIR29C
1q41	MIR194-1, MIR215
2p16.1	MIR216A, MIR216B, MIR217
3p21.1	MIRLET7G, MIR135A1
3p21.31	MIR191, MIR425
3q25.33	MIR15A, MIR16-1
4q25	MIR302A, MIR302B, MIR302C, MIR302D, MIR367
5q11.2	MIR449A, MIR449B
5q32	MIR143, MIR145
6p12.2	MIR133B, MIR206
6q13	MIR30A, MIR30C2
7q21.3	MIR489, MIR653
7q22.1	MIR25, MIR93, MIR106B
7q32.2	MIR96, MIR182, MIR183
7q32.3	MIR29A, MIR29B-1
8q22.2	MIR599, MIR875
8q24.22	MIR30B, MIR30D
8q24.3	MIR939, MIR1234
9p21.1	MIR873, MIR876
9q22.32	MIRLET7A1, MIRLET7D, MIRLET7F1
9q22.32	MIR23B, MIR24-1, MIR27B
9q33.3	MIR181A2, MIR181B2
11q13.1	MIR192, MIR194-2
11q23.1	MIR34B, MIR34C
11q24.1	MIRLET7A2, MIR100, MIR125B1
12p13.31	MIR141, MIR200C
12q13.13	MIR196A2, MIR615
13q14.3	MIR15A, MIR16-1
13q31.3	MIR17, MIR18A, MIR19A, MIR19B-1, MIR20A, MIR92A1
14q11.2	MIR208A, MIR208B

Locus	Cluster members
14q32.31	MIR127, MIR136, MIR337, MIR370, MIR431, MIR432, MIR433, MIR493, MIR665, MIR770
14q32.31	MIR134, MIR154, MIR299, MIR300, MIR323, MIR329-1, MIR329-2, MIR369, MIR376A1, MIR376A2, MIR376B, MIR376C, MIR377, MIR379, MIR380, MIR381, MIR382, MIR409, MIR410, MIR411, MIR412, MIR453, MIR485, MIR487A, MIR487B, MIR494, MIR495, MIR496, MIR539, MIR541, MIR543, MIR544, MIR654, MIR655, MIR656, MIR668, MIR758, MIR889, MIR1185-1, MIR1185-2, MIR1197
15q26.1	MIR7-2, MIR1179
16p13.12	MIR193B, MIR365-1
17p13.1	MIR195, MIR497
17p13.3	MIR132, MIR212
17q11.2	MIR144, MIR451
17q11.2	MIR193A, MIR365-2
17q22	MIR301A, MIR454
17q25.3	MIR338, MIR657, MIR1250
18q11.2	MIR1-2, MIR133A1
19p13.12	MIR181C, MIR181D, MIR23A, MIR24-2, MIR27A
19q13.32	MIR330, MIR642
19q13.33	MIRLET7E, MIR125A, MIR99B
19q13.42	MIR371, MIR372, MIR373, MIR498, MIR512-1, MIR512-2, MIR515-1, MIR515-2, MIR516A1, MIR516A2, MIR516B1, MIR516B2, MIR517A, MIR517B, MIR517C, MIR518A1, MIR518A2, MIR518B, MIR518C, MIR518D, MIR518E, MIR518F, MIR519A1, MIR519A2, MIR519B, MIR519C, MIR519D, MIR519E, MIR520A, MIR520B, MIR520C, MIR520D, MIR520E, MIR520F, MIR520G, MIR520H, MIR521-1, MIR521-2, MIR522, MIR523, MIR524, MIR525, MIR526A1, MIR526A2, MIR526B MIR527, MIR1283-1, MIR1283-2, MIR1323
20q13.32	MIR296, MIR298
20q13.33	MIR1-1, MIR133A2
20q13.33	MIR647, MIR941-1, MIR941-2, MIR941-3, MIR941-4, MIR1914
20q13.33	MIR941-1, MIR941-2, MIR941-3
21q21.1	MIRLET7C, MIR99A
22q11.21	MIR130B, MIR301B
22q13.1	MIR658, MIR659
22q13.31	MIRLET7A3, MIRLET7B
Xp11.22	MIRLET7F2, MIR98
Xp11.23	MIR188, MIR362, MIR500, MIR501, MIR502, MIR532, MIR660
Xp11.3	MIR221, MIR222
Xq13.2	MIR374A, MIR545
Xq13.2	MIR374B, MIR421
Xq23	MIR1264, MIR1912
Xq26.2	MIR18B, MIR19B-2, MIR20B, MIR92A2, MIR106A, MIR363
Xq26.3	MIR424, MIR450A1, MIR450A2, MIR450B, MIR503, MIR542
Xq27.3	MIR506, MIR507, MIR508, MIR509-1, MIR509-2, MIR509-3, MIR510, MIR513A1, MIR513A2, MIR514-1, MIR514-2, MIR514-3
Xq27.3	MIR888, MIR890, MIR891A, MIR891B, MIR892A, MIR892B
Xq28	MIR105-1, MIR105-2, MIR767
Xq28	MIR224, MIR452

INDEX

A

aberrant, 30, 96, 115
accuracy, 109
ACTH, 32, 53, 83
activation, 26, 29, 38, 54, 67, 111
adenocarcinoma, 30, 98
adenomas, 32, 84
adhesion, 108
adipose, 97
adipose tissue, 97
adult, 24, 30
age, 55, 100
aggressiveness, 28, 51, 69, 80, 88
algae, 3
algorithm, 12, 18
alternative, 110
amine, 10
angiogenesis, 33, 87, 93, 112
angiogenic, 32, 40, 99
animals, 3
annexin, 61, 96
annotation, 2, 4, 20
antagonist, 70, 105
antiapoptotic, 36
antibody, 108
anticancer, 67, 111, 112
antigen, 82, 99
antisense, 4, 38, 41

apoptosis, 27, 30, 32, 33, 36, 38, 45, 49, 53, 54, 61, 69, 72, 84, 86, 103
apoptotic, 31, 40, 54, 59, 111, 112
apoptotic effect, 54, 111
apoptotic pathway, 31
application, 9, 14, 20, 53, 73
artificial, 43, 51, 58, 60, 71, 73, 74, 76, 77, 78, 80, 81
assessment, 97, 107, 108
associations, 36, 55
atypical, 39, 53, 111

B

background noise, 12
B-cell, 103
behavior, 36, 108
benefits, 17
beta, 65
binding, 1, 2, 10, 25, 34, 38, 49, 53, 64, 65, 70, 114
biogenesis, 1, 2, 109
bioinformatics, 18, 19, 46, 48, 53, 60
biological, xiii, 41, 71, 98
biologically, 32, 92
biology, xiii, xiv, 1, 94, 114
biomarkers, 3, 98, 107, 109, 113
biopsies, 97
biopsy, 109
blood, 69, 115

blot, 15, 66, 114
BMPs, 63
bonds, 11
bone, 37, 42, 50, 63, 69, 76
bone morphogenetic proteins, 63
branching, 3
BRCA1, 55
BRCA2, 55
Breast, i, iii, iv, v, vii, xi, xiv, 23, 75, 87, 89, 90, 91, 92, 94, 96, 97, 99, 107, 113, 114, 115
breast carcinoma, 31
bypass, 97

C

cadherin, 63, 65
Caenorhabditis elegans, 3, 24
Cancer, i, ii, iii, iv, v, vii, xi, xiv, 4, 5, 14, 23, 83, 85, 86, 87, 88, 89, 90, 91, 92, 93, 94, 95, 96, 97, 98, 99, 100, 102, 104, 105, 107, 113, 114, 115, 116
cancer cells, 11, 29, 30, 33, 42, 50, 51, 56, 58, 59, 64, 65, 71, 82, 85, 91, 99, 103, 104, 108, 116
cancer progression, 41, 42, 63, 99
cancer screening, 110
cancer stem cells, 25, 26
capacity, 25, 54, 62, 76
carcinogenesis, 39
carcinoma, 39, 53, 64, 86, 99, 111
cardiovascular disease, 57
caspase, 27, 102
CD44, 38, 77, 84
CDK, 71
CDK6, 24, 56
cDNA, 8, 10, 57
cell culture, 76
cell cycle, 24, 29, 32, 40, 50, 59, 62, 70, 73, 80, 86, 88, 102
cell death, 29, 37, 41, 45, 59, 73, 88, 96, 102
cell growth, 30, 36, 41, 43, 49, 50, 53, 54, 58, 72, 89, 103, 104, 105
cell invasion, 36, 41, 60, 66

cell lines, 26, 27, 29, 34, 40, 58, 61, 77, 82, 83, 86, 88, 93, 95, 96
cervical, 50, 57, 103
cervical cancer, 50, 103
chemoresistance, 39, 108
chemotherapeutic agent, 65, 111
chemotherapy, 25, 39, 79, 84, 86, 95, 111, 112, 115
Chinese women, 90
chromatin, 25, 29, 34
chromosomal instability, 29, 80
Chromosomes, 89
chronic, 32, 57, 88, 103
chronic lymphocytic leukemia, 32, 57, 88
circulation, 69
classification, 2, 14, 103
cleavage, 75
clinical, xiv, 31, 36, 37, 53, 68, 76, 77, 97, 104, 107, 109, 112, 116
clinicopathologic, 101
cloning, 7, 9
clustering, 20
clusters, 1, 3, 33, 40, 46, 47, 59, 62, 64, 71, 92, 102, 139
c-Myc, 26, 38, 45, 100
Cochrane, 63, 65, 86
coding, 1, 24, 27, 33, 54, 75, 94, 111
cohort, 44, 76
colon, 33, 36, 47, 50, 89, 100, 109, 110
colonization, 44
colorectal, 53, 63, 101
compilation, 17, 18
complementarity, 4
complementary, 1, 8, 11, 12, 54, 111
complementary DNA, 8
complexity, 45, 108
complications, 108
components, 2, 34
concentration, 39
consensus, 82
conservation, 18
control, 1, 24, 30, 54, 61, 65, 79, 84, 87, 91, 105, 112, 113
controlled, 65, 110
corepressor, 91

correlation, 7, 29, 61, 107
cross-linking, 15
cyclin D1, 24, 34, 105
cytokine, 51
cytometry, 50
cytoplasm, 2
cytosol, 38

D

data set, 101
database, 19, 20, 21
death, 54, 101, 111, 116
definition, 108
degradation, 2, 39
degree, 24
delivery, 111
density, 109
deregulation, 25, 38, 91, 98, 114
detection, 8, 9, 11, 12, 13, 14, 15, 53, 109, 115
diagnostic, 86, 98, 107, 108
diagnostic markers, 108
differentiation, 10, 24, 57, 63, 97, 102
disease progression, 37
disease-free survival, 36, 99
diseases, 24, 48, 80
disposition, 75
distal, 50, 76
distribution, 38, 52, 67
diversity, 31, 108
DNA, 8, 10, 30, 46, 56, 87, 92, 96, 111, 114
downregulating, 54
down-regulation, 27, 32, 35, 36, 37, 40, 53, 56, 73, 75, 76, 79, 84, 111
Drosophila, 19, 24
drug resistance, xiii, 79, 112
drug treatment, 56
drug-induced, 27
dsRNA, 2

E

E2F, 33, 70, 83, 101, 104
E-cadherin, 42, 63, 65, 85, 91, 93
efficacy, 108
EGF, 68
EGFR, 29, 55, 68
electrocatalytic, 11, 14
electrochemical detection, 7
electronic, ix
electrophoresis, 8, 9
electrostatic, ix
embryonic, 15, 24, 25, 50
emission, 10
EMT, xiii, 25, 57, 63, 64, 65, 66, 85, 91, 113
encoding, 32, 65, 68
endocrine, 72
endogenous, 29, 37, 41, 59, 77, 109
entropy, 20
enzymes, 40, 59
epidermal, 29, 83, 92, 103
epidermal growth factor receptor, 29, 83, 92, 103
epigenetic, 26, 30, 56, 90
epigenetic mechanism, 30
epigenetic silencing, 31
epithelia, 27
epithelial cell, 25, 31, 38, 39, 57, 64, 90, 93, 100
epithelial cells, 25, 31, 38, 39, 57, 64, 90
epithelial ovarian cancer, 88
ERK1, 29
esophageal adenocarcinoma, 61
estradiol, 26, 84
estrogen, xiii, xiv, 23, 26, 27, 30, 35, 39, 43, 45, 58, 68, 84, 85, 88, 91, 93, 95, 99, 100, 101, 105, 116
Estrogen, 97, 114
evidence, 25, 34, 89
evolution, 101, 108
evolutionary, 31
excitation, 12
exogenous, 45, 65
exposure, 30
extracellular, 29, 76
extracellular matrix, 76
extravasation, 44

F

fabricate, 11
familial, 55, 100
family, xiii, xv, 2, 3, 24, 25, 26, 27, 33, 35, 37, 38, 43, 44, 45, 48, 52, 57, 62, 63, 64, 65, 66, 67, 71, 78, 85, 89, 93, 99, 102, 110, 111, 112, 114
family members, xiii, 24, 26, 35, 38, 43, 44, 45, 63, 64, 65, 66, 71, 110, 112
fat, 69
F-box, 25
feedback, 27, 34, 35, 64, 68, 83, 85, 101, 105
fixation, 8
fluorescence, 7, 10, 113
fluorophores, 12
fractionation, 7
functional analysis, 55, 90

G

gastric, 32, 47, 92, 104
gel, 8, 9
gene, 1, 2, 3, 4, 5, 9, 17, 18, 19, 20, 21, 30, 31, 32, 33, 34, 37, 41, 43, 46, 48, 49, 58, 61, 65, 79, 82, 84, 87, 91, 94, 95, 96, 98, 100, 101, 102, 103, 104, 105, 108, 110, 112, 114, 115, 116, 119
gene amplification, 48
gene expression, 3, 9, 19, 35, 87, 91, 98, 101, 102, 104, 108, 114
genes, 1, 3, 4, 17, 18, 19, 20, 25, 31, 32, 34, 36, 37, 40, 43, 44, 53, 55, 56, 58, 59, 61, 65, 69, 76, 89, 91, 94, 95, 97, 102, 105, 108, 110
genetic, 61, 77, 90
genetics, 94, 114
genome, 1, 18, 62
genomic, 19, 20, 109
genomics, 4, 19
gland, 57
glass, 10, 13
glioblastoma, 29, 92
gold, 7, 13

gold standard, 7, 13
growth, 24, 29, 35, 36, 37, 49, 51, 53, 60, 61, 64, 66, 82, 89, 96, 97, 102, 111
growth factors, 37
growth inhibition, 36
guardian, 89
guidelines, 2

H

hematopoietic, 57
hepatitis C, 102
hepatocellular, 50, 102, 104
hepatocellular carcinoma, 50, 102, 104
HER2, xiii, 23, 31, 36, 43, 48, 58, 67, 69, 72, 76, 77, 81, 90, 96, 107, 111, 113
HER3, 91
heterochromatic, 133
heterogeneous, 25, 108
high-risk populations, 110
histological, 37, 103
histone, 34
HNF, 94
homology, 70
hormone, 36, 108
human, xiii, 2, 3, 4, 11, 14, 19, 20, 21, 24, 25, 27, 29, 30, 41, 44, 48, 49, 50, 55, 56, 59, 64, 65, 68, 72, 75, 82, 83, 84, 85, 86, 87, 88, 91, 93, 95, 96, 97, 98, 99, 101, 102, 103, 104, 109, 114, 115, 116, 119
human estrogen receptor, 83
human genome, 3
humans, 17, 75
hybridization, 8, 11, 13, 24, 56
hydrazine, 12
hypermethylation, 30, 47, 55, 56, 82, 96
hyperplasia, 53, 111
hypothesis, 25, 63, 65
hypoxia, xiii, 34, 42, 46, 70, 85, 93, 105, 112
hypoxic, 40, 58, 87

I

identification, 4, 19, 20, 21, 109

identity, 10
IGF, 50
IGF-I, 50
IGF-IR, 50
IL-1, 54
immunity, 57
immunodeficient, 25, 60, 69
in situ, 7, 10, 27, 38, 39, 52, 67, 99, 111
in situ hybridization, 7, 10, 27, 38, 52, 67
in vitro, 26, 30, 36, 53, 55, 64, 70, 76, 77, 92, 96
in vivo, 11, 12, 50, 53, 64, 76, 77, 92, 96
inactivation, 48, 56, 92, 95
inactive, 67
incurable, 113
indicators, 4, 108, 109
indolent, 42
inducer, 63
induction, 25, 27, 41, 44, 59, 65, 70, 100, 111
infection, 48
inflammation, 27, 57
inflammatory, 26, 32
inhibition, 26, 29, 34, 37, 39, 41, 44, 53, 59, 97, 101
inhibitors, 34, 38, 41, 59, 67, 71, 92, 112
inhibitory, 39, 68, 69
initiation, 39
injuries, 30
insight, xiii, 38
institutions, xiv
insulin-like growth factor, 32, 35, 50
integrin, 25, 44
intensity, 10
interaction, 54, 67, 82, 84, 111
interactions, 34
interference, 4, 5
interleukin, 54
international, 2
interpretation, 17, 18
intervention, 57
intron, 33, 46
invasive, 25, 29, 39, 57, 63, 76, 82, 99, 109
IRS, 50, 105
island, 31, 62, 96

isoforms, 65, 91

J

JT, 14, 87, 89, 97, 100
Jun, 14, 20, 85, 87, 88, 92, 94, 96, 97, 98, 99, 102, 114, 115, 116
Jung, 57, 92

K

kidney, 50
kinase, 29, 30, 54, 67, 68, 71, 89, 99, 112

L

labeling, 14
large-scale, 36
lead, 112
learning, 20
lesions, 31, 42, 43, 51, 52, 53, 73, 76, 111
leukaemia, 103
leukemia, 47, 85, 87
ligand, 114
links, 88, 92, 100
lipid, 46, 104
lipid metabolism, 104
literature, 47
liver, 53
localization, 49
location, 10
locus, 33
loss of heterozygosity, 33
low-density, 114
luciferase, 11, 37, 61, 66, 69, 75, 82
Luciferase, 75
luminal, 27, 28, 31, 38, 39, 43, 68, 76, 77, 83
lung, 29, 33, 36, 47, 50, 51, 55, 57, 66, 69, 70, 76, 86, 96, 105, 109, 110
lung cancer, 51, 96, 105
lymph node, 27, 88, 104, 108, 116
lymphoma, 57
lymphomagenesis, 100

M

machinery, 4
maintenance, 10, 41, 60
malignancy, 92
malignant, 27, 38, 50, 66
mammalian tissues, 15
mammals, 3, 21, 24, 33
management, 5, 116
MAPK, 68
mapping, 98
maternal, 113
maturation, 87
MDA, 54, 55, 60, 63, 66, 68, 69, 72, 76, 94, 98
MDR, 74
measures, 13
mechanical, ix
medulloblastoma, 30, 87
melanoma, 55, 60, 87, 94
membranes, 15
memory, 30, 42
mesenchymal, xiii, 25, 57, 63, 64, 65, 85, 89, 93, 113, 114
messenger RNA, xiii, 1, 2, 32, 55, 65, 83, 108
metabolic pathways, 46, 104
metabolism, 40, 59
metastases, 69
metastasis, xiii, xiv, 25, 30, 31, 32, 37, 44, 50, 55, 63, 65, 66, 69, 75, 77, 88, 90, 93, 96, 97, 101, 102, 104, 105, 107, 108, 114, 116
metastatic, 25, 31, 44, 50, 54, 55, 60, 65, 69, 75, 77, 82, 84, 99, 113
metastatic cancer, 55
metazoan, 19
methylation, 30, 44, 56, 96, 115
mice, 25, 69, 97
microarray, 8, 9, 10, 11, 15, 24, 37, 40, 56, 57, 58, 62, 71, 76, 79, 95, 109, 114, 115
Microarrays, 10
microcosm, 18
microenvironment, 70
microRNAs, 4, 5, 14, 15, 19, 20, 21, 30, 59, 84, 85, 89, 90, 92, 93, 95, 96, 97, 98, 99, 100, 101, 103, 113, 114, 115, 117
microtubule, 65, 86
migration, 31, 41, 49, 50, 54, 55, 57, 64, 65, 69, 76, 77, 93, 110
mitogen, 83
mitogen-activated protein kinase, 83
modules, 17, 18
moieties, 11
molecular markers, 108
molecules, 1, 3, 7, 9, 11, 12
morphological, 31
morphology, 69
mouse, 3, 24, 26, 36, 60
mRNA, 3, 17, 18, 20, 29, 34, 35, 39, 40, 46, 48, 53, 55, 58, 59, 61, 68, 70, 72, 74, 75, 82, 87, 90, 99, 108, 109
multidrug resistance, 74, 79, 96, 104, 112, 115
multiple myeloma, 47
mutagen, 62
mutant, 62
mutation, 14, 49, 95
MYC, 24, 33, 53, 56, 70, 100, 105
myeloid, 58, 87
myeloid cells, 58, 87

N

nanorods, 13
nematode, 3
neoplasia, 89
neoplastic cells, 89
network, 45, 62, 70, 83, 89
neural networks, 43, 52, 58, 60, 71, 73, 74, 76, 77, 78, 80, 81
neuroblastoma, 47
neuroendocrine, 47
neuronal cells, 65
NF-κB, 26, 54
normal, 10, 11, 13, 23, 27, 28, 29, 30, 31, 36, 38, 39, 40, 43, 44, 45, 47, 48, 51, 52, 54, 56, 57, 58, 59, 61, 64, 66, 67, 69, 70, 73, 74, 77, 79, 100, 110, 111

normal stem cell, 59, 64, 100
nuclear, 2, 34
nucleic acid, 8, 12, 15, 110
nucleotides, 1, 2
nucleus, 38

O

observations, 36, 73
oestrogen, 89, 94, 96
oestrogen receptor-positive breast cancer, 89
oligonucleotides, 8, 12, 110
oncogene, 32, 33, 34, 36, 50, 56
Oncogene, 84, 85, 92, 95, 96, 97, 98, 101, 116
oncogenes, xiii, 23, 27, 33, 45, 107, 110
oncogenesis, 39, 65
oncology, 101, 107
organic, 12
organization, 20
ovarian, 30, 53, 55, 91, 94, 100, 109
ovarian cancer, 30, 55, 91, 94, 100
oxidation, 12
oxygen, 42, 46

P

p53, 31, 44, 53, 89, 90, 94, 100, 112
pairing, 20, 95
pancreas, 33, 47, 54, 57, 58
pancreatic, 30, 36, 47, 63, 89, 95, 98, 110
paraffin-embedded, 38, 52, 67, 113, 114, 115, 116
pathology, 94, 114
pathways, 111
patients, 32, 36, 44, 50, 54, 55, 61, 76, 108, 110, 111, 112
PCR, 7, 9, 11, 14, 49, 60, 68
penetrance, 26
pharmacokinetic, 75
phenotype, 57, 63, 68, 77, 83
phosphorylation, 38, 67
physiological, 24, 57
PI3K, 67

pituitary, 32, 53, 83, 84
pituitary tumors, 32, 83
placental, 113
plasma, 107, 109, 113
plasmid, 53, 61, 75
plasminogen, 60, 95
plasticity, 93
play, xiii, 1, 26, 57, 63, 64, 65, 71, 72, 74
pluripotency, 99
polyacrylamide, 8, 9
polycomb group, 85, 100
polymerase, 2
polymorphism, 55, 100
poor, 10, 32, 36, 50, 76, 99, 104, 111, 116
population, 59, 61, 64, 79
positive feedback, 27
precursor cells, 25
prediction, 3, 17, 18, 20, 21, 55, 104
predictive marker, 109
predictors, 86
preparation, ix
prevention, 111
primary tumor, 69, 113
probe, 8, 10, 11, 12
procedures, 9
production, 38, 55, 84
progenitor cells, 30, 59, 64
progeny, 30
progesterone, 27, 43, 58, 78, 80, 96, 110
prognosis, 4, 32, 37, 68, 79, 104, 105, 110, 117
prognostic marker, 37, 49, 109
proliferation, 24, 28, 31, 33, 34, 35, 36, 39, 41, 42, 43, 49, 50, 52, 54, 56, 58, 61, 66, 69, 71, 72, 88, 90, 95, 103, 105, 108, 110
promoter, 31, 32, 35, 44, 53, 57, 64, 82
promyelocytic, 87
prostate, 32, 33, 36, 47, 53, 54, 58, 84, 88, 97, 98, 109, 110
prostate cancer, 84, 98
prostate carcinoma, 88, 97
protein, 2, 25, 29, 37, 38, 40, 41, 42, 48, 49, 50, 53, 54, 57, 59, 60, 61, 65, 66, 70, 72, 73, 74, 75, 80, 82, 83, 84, 85, 96, 98, 99, 100, 102, 111, 115, 116

protein kinase C (PKC), 38
protein kinases, 65
proteins, 2, 26, 38, 45, 48, 54, 68, 87, 90
proteolysis, 108
proteomics, 19
protocols, 9
proto-oncogene, 53
proximal, 64
PSA, iv
public, 19

R

radiation, 45
radiolabeled, 8
Raman, 7, 12, 13, 14
Raman spectroscopy, 7, 13, 14
range, 12, 74
RAS, 24, 27
REA, 91
reading, 5
reality, 108
real-time, 7, 9, 60
receptor-positive, 88
receptors, 63
recognition, 1, 2, 4, 37, 48, 61
redox, 11
reduction, 38, 39, 76, 84, 111
regression, 37
regression analysis, 37
regulation, 1, 3, 17, 18, 21, 32, 35, 36, 40, 41, 44, 52, 58, 59, 63, 64, 71, 72, 74, 77, 83, 84, 87, 88, 90, 104
regulators, 24, 50, 72, 75
relapse, 50, 76, 79
relevance, 108
remodeling, 25, 34
renal, 57, 92
repair, 70
replication, 46
reporters, 37
repression, 3, 44, 49, 53, 58, 63, 71, 85, 90
repressor, 42, 70, 100
research, 3, 104

resistance, 4, 39, 65, 67, 74, 75, 79, 84, 93, 95, 98, 99, 105, 114, 115, 116
responsiveness, 67, 108, 109, 112
restoration, 50, 96, 112
Rho, 53
ribonucleic acid, 83
ribose, 8
RISC, 2
risk, 30, 62, 76, 90
risks, 61, 79
RNA, 1, 2, 4, 5, 7, 8, 9, 10, 11, 14, 15, 19, 20, 25, 38, 49, 68, 89, 93, 94, 100, 104, 114, 115, 116, 117
RNAi, 5
RNAs, xiii, 1, 4, 19, 92, 94, 101

S

S phase, 86
sample, 11, 12, 13
scaffold, 54
scientific, xiv
search, 17, 18, 19, 48, 50, 55, 60, 72, 75
secretion, 32
seed, 1, 2, 20, 24, 25, 45, 48, 95
segmentation, 10
self-renewal, 25, 64
self-renewing, 26
sensitivity, 8, 9, 11, 12, 13, 65, 75, 79, 86, 108, 112
sequencing, 9, 14, 94
series, 10, 17, 30, 47, 55, 56, 63, 82
serine, 65
serum, 11, 14, 107, 109, 111, 113, 115
signaling, 23, 29, 39, 43, 48, 51, 54, 58, 68, 70, 80, 83, 84, 89, 95, 99, 103, 111
signaling pathway, 23, 54
signaling pathways, 23
silver, 13
similarity, 24
single nucleotide polymorphism, 61, 79
siRNA, 15, 37, 41, 53, 59
sites, 9, 25, 29, 39, 41, 54, 59, 68, 70, 111
SNP, 61, 79
SNPs, 79

software, 18
solid tumors, 31, 65, 103, 116
spatial, 38, 52, 67
species, 2, 18, 110
specificity, 13, 40, 98
spectra, 13
spectroscopy, 7
spectrum, 40, 42, 46
speed, 9
stability, xiii, 1, 2
stages, 10, 71
standardization, 7, 13
standards, 13
stem cells, 64, 85, 86, 97, 100
steroid, 63, 68
stomach, 33, 36, 110
strategies, 79, 108, 110
streptavidin, 10
stress, 49
substrates, 86
Sun, 100, 104, 105
suppression, 24, 37, 40, 44, 48, 49, 77, 92, 100, 102, 111, 116
suppressor, xiii, 30, 34, 35, 38, 40, 41, 47, 49, 50, 51, 53, 56, 59, 67, 76, 84, 89, 90, 95, 96, 100, 105, 110, 111
suppressors, 23, 32, 37, 107, 110, 111
survival, 36, 40, 44, 45, 50, 53, 67, 76, 97, 108
susceptibility, 61, 62, 80
syndrome, 2
synthetic, 30

T

tamoxifen, 37, 40, 58, 71, 72, 73, 76, 79, 98, 105
targets, 3, 4, 12, 18, 19, 20, 24, 34, 37, 42, 44, 45, 46, 61, 65, 67, 72, 75, 83, 95, 96, 97, 99, 101, 103, 105, 107, 109, 110, 112, 116
T-cell, 102
technology, 10, 113
temperature, 8
tenascin, 76
TGF, 25, 36, 37, 42, 57, 63, 66, 93, 99
therapeutic, 4, 36, 55, 57, 74, 79, 82, 86, 108, 109, 110, 113
therapeutic agents, 74, 86
therapeutic targets, 5, 109
therapeutics, 102
therapists, 115
therapy, 4, 25, 54, 58, 72, 108, 109, 110, 111, 112
thermal, 8
thermal stability, 8
three-dimensional, 4
threonine, 65
thyroid, 57, 86, 89, 98, 102
thyroid carcinoma, 86, 89
tight junction, 57
time, 8, 9, 11, 13, 14, 30, 49
tissue, 10, 23, 25, 27, 28, 29, 31, 36, 38, 40, 43, 44, 47, 48, 52, 56, 59, 66, 67, 73, 92, 111
TNC, 76
TNF, 54
Toll-like, 54
TP53, 54, 101, 111, 116
trans, 67, 77
transcript, 2, 33, 48, 64
transcription, 3, 5, 9, 10, 25, 26, 29, 31, 33, 34, 35, 40, 45, 59, 63, 64, 68, 70, 76, 82, 91, 97, 98, 104
transcription factor, 5, 25, 27, 29, 31, 33, 34, 35, 40, 45, 59, 63, 68, 70, 76, 82, 91, 97, 98, 104
transcription factors, 33, 34, 45, 98, 104
transcriptional, 27, 31, 42, 44, 63, 70, 85, 93
transcriptomics, 19
transcripts, 1, 33
transfection, 36, 75
transfer, 8
transformation, 27, 34, 41, 59, 96
transforming growth factor, 25
transition, xiii, 25, 57, 63, 85, 86, 89, 93, 113, 114
translation, 2, 32, 34, 39, 50, 58, 87, 90
translational, xiii, 1, 3, 34, 39
trastuzumab, 108, 112

tumor, xiii, 23, 24, 25, 27, 30, 32, 33, 34, 35, 36, 37, 38, 40, 44, 47, 49, 50, 51, 52, 53, 54, 56, 58, 59, 61, 64, 67, 69, 70, 71, 74, 76, 77, 79, 82, 84, 87, 89, 95, 96, 99, 100, 101, 103, 105, 107, 108, 110, 111, 112, 116
tumor growth, 36, 50, 56, 101, 110, 116
tumor progression, 95
tumorigenesis, 34, 36, 42, 49, 62, 67, 86, 102
tumorigenic, 29
tyrosine, 30, 66, 67, 112

U

ubiquitin, 25
uniform, 4
urokinase, 60, 95
UTRs, 21

V

variables, 109
variation, 47

vascular, 30, 31, 40, 43, 66
vector, 26, 38, 72
VEGF, 40, 66, 70, 96
viral, 111
virus, 102
vitamin D, 48

W

Weinberg, 97, 102
Wnt signaling, 100
women, 58, 110
worms, 2, 3

X

xenograft, 25, 36, 56, 60

Z

zebrafish, 10, 15
zinc, 87